The
MONSOON
LANDS of ASIA

The MONSOON LANDS of ASIA

R.R.Rawson

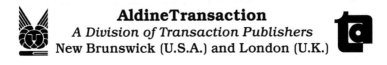

AldineTransaction
A Division of Transaction Publishers
New Brunswick (U.S.A.) and London (U.K.)

First paperback printing 2007
Copyright © 1963 by R. R. Rawson.

This book is printed on acid-free paper that meets the American National Standard for Permanence of Paper for Printed Library Materials.

Library of Congress Catalog Number: 200605188
ISBN: 978-0-202-30942-2
Printed in the United States of America

Library of Congress Cataloging-in-Publication Data

Rawson, R. R. (Robert Rees)
    The monsoon lands of Asia / R. R. Rawson.
        p. cm.
    Reprint. Originally published: Chicago : Aldine Pub. Co., 1963.
    Includes bibliographical references and index.
    ISBN 978-0-202-30942-2 (pbk. : alk. paper)
      1. Asia—Geography. I. Title.

DS5.92.R38   2007
915—dc22

                                                          2006051888

Contents

	List of Plates	7
	List of Figures	9
	List of Tables	11
	Preface	13
1	Introduction	15
2	Structure and Relief	22
3	Climate	33
4	Natural Vegetation	47
5	Soils	51
6	Historical Geography	54
7	Agriculture	69
8	Manufacturing Industry	80
9	India: the British Period and the Partition of 1947	85
10	The Republic of India	101
11	Pakistan and Kashmir	128
12	Ceylon	148
13	China	164
14	Japan and Korea	203
15	South-east Asia	216
16	Postscript	243
	Selected Bibliography	245
	Index	248

List of Plates

1   A crowded bazaar at Agra   *facing page* 48

2   Moghul architecture in India: the Taj Mahal at Agra   49

3   Moghul architecture in India: the Red Fort at Delhi   64

4   Modern architecture in India: Marine Drive, Bombay   65

5   Foothills of the eastern Himalayas: a farmstead and a flight of
    paddy terraces on a steep hillside   96

6   December scene on the Bengal plain, near Comilla   97

7   Rourkela: one of the newer iron and steel plants of India   112

8   A multi-purpose hydro-electric scheme in India: the Hirakud
    dam on the Mahanadi river   113

9   Part of the Himalayan frontier of India: Kanchenjunga, and the
    jungly ranges of Sikkim and Nepal   144

10 Difficulties of road building in the Himalayas: a landslip in a
    gorge on the Kalimpong-Lhasa trade route   145

11 The Great Wall of China, north of Peking   160

12 Hong Kong: Kai Tak airport   161

13 Grassland and livestock in Sinkiang   192

14 Malaya: aerial view of a tin dredge   193

15 Malaya: three dredges working tin-bearing alluvium near
    Kuala Lumpur   193

16 Malaya: a gravel-pump tin mine   208

17 A crowded quay on Singapore river, in the heart of the city   209

List of Figures

1   Monsoon Asia: structure   25
2   Geological section: Sumatra   31
3   Colombo: climatic data for 1938   36–7
4   Monsoon Asia: tracks of typhoons and winter depressions   39
5   Monsoon Asia: average annual rainfall   40
6   The Indus civilization and the Mauryan empire   57
7   The Gupta empire   59
8   Vijayanagar, and the Moghul empire   62
9   The Han empire   64
10 The Tang and Sung empires   65
11 India in the late nineteenth century   89
12 India: railways   91
13 India: the cotton industry   93
14 India: the iron and steel industry   96
15 The jute-growing areas of India and Pakistan   99
16 Five of the thirteen cities of Delhi   106
17 *Bhangar, khadar,* bluffs and settlement on the Ganges plain
        30 miles south-east of Meerut   107
18 The Ganges plains   109
19 The Bhakra-Nangal scheme   111
20 The Ganges Canal headworks   113
21 Central India and the peninsula   115
22 The Damodar valley scheme   117
23 Disputed territory on the Himalayan border   126
24 The lower Indus valley and part of Baluchistan   130

25	Punjab irrigation: the Triple Canal Project	133
26	West Pakistan: the north-west hills area	137
27	East Pakistan	140
28	Kashmir	145
29	Ceylon	150
30	Ceylon: Minneriya tank and old irrigation channels	159
31	China: coalfields, and the iron and steel industry	166–7
32	The Hwang Ho development scheme	169
33	China: railways (1)	172
34	China: railways (2)	173
35	China: agricultural regions	177
36	The North China plain	180
37	The Szechwan Basin and the middle Yangtze	186
38	Tibet, Sinkiang and Mongolia	191
39	Manchuria	194
40	Hong Kong	201
41	Japan	205
42	Burma	221
43	Arakan coast: mangrove reclamation	224
44	Malaya	227
45	Thailand and Indo-China	230
46	Java	235
47	Sumatra	237
48	Sumatra: detail of oilfield in mangrove swamp	238

List of Tables

1   Monsoon Asia: average rainfall   41

2   Indices of food production per head of population   73

3   Crop yields: 1962   75

4   Coal and electricity production per head of population: 1960   82

5   India: benefits expected from hydro-electric projects   103

6   Ceylon: rainfall in 1938, with offsets from the average for the years 1911–30   153

7   Japan: distribution of foreign trade   211

Preface

This book is an attempt to explain the geography of Monsoon Asia to first-year university students and sixth-form pupils. In the early stages of the work I received a great deal of encouragement and advice from the late Mr G. J. Cons, M.A. The statistics relating to subdivision of land on page 76 are reproduced by permission of Dr A. Sharan, University of Patna. Miss Ann Howgate read the text, and suggested a host of improvements in expression which I gratefully adopted. I am also indebted to Mrs S. M. Weston who advised on the drafting of the maps and prepared the final drawings.

For the defects of the book, I alone am responsible.

R. R. RAWSON

I

Introduction

Monsoon Asia is the southern and eastern part of Asia, including India, Pakistan and Ceylon; China and Japan; and all the peninsulas and islands of the south-east as far as New Guinea and the Philippines. But it is more than an area of land or an assemblage of countries. It is a geographical region. To be acceptable as a geographical region an area must satisfy the following conditions:

1—however striking the diversity of the physical setting and the languages, religions and general appearance of the people, there must be some aspect of culture (for example, the social outlook or the organization of the economy) which pervades the area in such a way as to justify the recognition and study of that area as one entity;
2—there must be substantial cultural differences between it and adjacent areas.

If the first of those conditions is applied to Monsoon Asia the diversity is at once apparent. There are enormous climatic differences, for example, between southern Malaya, where the mean temperature never drops below 80°F. and rain falls throughout the year, and northern China, where monthly mean temperatures are below freezing for three months and the rainfall is concentrated into four summer months. Selecting two areas on the same latitude, the aridity of the Indus lowlands contrasts sharply with the humid conditions of Bengal. Associated with the climatic differences are the obvious differences in wild vegetation and in the field crops.

Equally impressive is the variety in things which have no direct relationship with the physical geography. In India, which apart from small islands in the Bay of Bengal is one continuous land area, there are six major languages, each the only tongue of several million people, not to mention scores of local dialects. Indians from the Madras countryside cannot converse with those of the Ganges plains. The 51 million people of East Pakistan speak Bengali; their 43 million compatriots a thousand miles away in West Pakistan speak Urdu. In China, Mandarin is spoken in the north and is the official language of the whole country. Yet in the south it is only slowly making headway against local languages such as Cantonese and Fukienese. The province of Fukien alone has over a hundred dialects. The number and distribution of major religions amount at first sight to confusion. China has Confucianism, Buddhism and Taoism; India has Hinduism, Islam and Buddhism. Burma and Siam are principally Buddhist, while in Malaya and Indonesia Muslims predominate but there are many Hindus. These things, together with other aspects of culture, such as costume, architecture and the detail of social organization, leave one in no doubt that Monsoon Asia is an area of rich diversity.

What aspect of culture is common to the whole? Material poverty. Almost everywhere the people are poor, in the sense that they have little to eat or to spend. Many are seriously underfed, and for the majority the expectation of life at birth is only about 35 years. Roughly 80 per cent live not in towns but in the countryside. They cultivate holdings which are mostly too small for a decent living. Some own their land, some are tenants, probably a quarter are labourers with no land of their own. In recent decades this condition has grown worse because everywhere the population has increased. There has been no commensurate increase either in food production from land already under cultivation or in the total area of farmland. Often where hardship is greatest no good virgin land remains. Two hundred years ago life was hard in rural areas in Europe, but there were opportunities for people to move to the towns and earn better livings in the factories. That solution was possible because capital had accumulated from commerce and was available to build the factories and establish the industries. For the most part, the countries of Monsoon Asia have very little capital in relation to their enormous populations, and manufacturing industries have made only slight progress. Some parts are better off than others. Japan, the most industrialized, has a full range of modern industry and is today the greatest shipbuilding country in the world. But even there poverty is a problem and nearly half the people are small farmers or farm labourers. Compare the United Kingdom, where only one in 20 lives by farming. Material poverty, a predominantly rural population and the urgent need of change in the economy to improve the lives of the people are characteristic of all the countries of Monsoon Asia: they, more than any other considerations, make the area a geographical entity.

Turning to the second condition, there is the question of whether Monsoon Asia is culturally distinct from neighbouring areas. To the north is the thinly populated Asian portion of U.S.S.R., Communist in organization and with its economic life dominated by influences from the longer-settled and more heavily populated European part of Russia. To the west is the arid and sparsely populated Middle East, a cultural entity characterized by Islam, pan-Arabism, the great oil interests and the Jewish ambitions for Israel. Monsoon Asia is culturally distinct from both, and separated from them by the great physical barrier of High Asia, a belt of largely uninhabited plateau and desert up to 2,000 miles wide extending from Persia to Manchuria. This latter fact is apparent from the small-scale atlas maps showing the distribution of world population. But some of those maps are deceptive in that they give the impression of continuous high density over great tracts of Monsoon Asia. Maps of larger scale show the true situation: several areas of very high population-density (e.g. the Ganges plains, the lowlands of China, the deltas of South-east Asia) separated by upland country with comparatively few people.

Monsoon Asia with its 1,500 million people, fully half the population of the entire world, is therefore acceptable as a geographical region. But it is a mistake to regard it as a museum-piece, a region where human affairs are fossilized in forms established long ago. Change is apparent everywhere. On the one hand there is a tendency for things to get worse. Each year there are more people to share the food. In crowded India alone the natural increase is nine million a year. On the other hand the governments of Monsoon Asia are doing much to improve conditions. In that task they are helped by international agencies such as the Colombo Plan, the World Bank, United Nations and also by individual states outside the area. This help has been forthcoming in the form of shipments of grain and fertilizers, capital grants and loans, technical advice and assistance in building dams and factories, and the training of Asian students in European, American and Russian colleges. Progress is slow but there are signs of headway. Thus 1955 was the first year since World War II in which the increase of home-grown food kept pace with the rise in population, but in 1957 the food production per head of population was still 14 per cent below pre-war levels.

DISEASE

Until remedies were discovered and applied during the last 60 years, millions died from such diseases as dysentery, plague, cholera and malaria. Many more were, and substantial numbers still are, too weakened by these diseases to do much to improve their farming. Despite modern medicine and public health regulations there is always a risk of new outbreaks. Malaria caused high death rates until the years following World War II when the breeding of mosquitoes was progressively controlled by spraying houses and water surfaces with preparations of DDT. There has been a spectacular reduction in deaths from malaria wherever DDT has been used. Areas formerly considered uninhabitable because of malaria are being opened up. But it is too soon to claim the remedy as permanent. It is at least possible that malarial mosquitoes, resistant to DDT and formerly few in number, might multiply. The same argument applies to certain other insect-borne diseases.

NATIONALISM

The work of economic improvement goes on in a social atmosphere of strong nationalism. Before World War II much of Monsoon Asia belonged to western colonial empires. Malaya, British Borneo, Ceylon and Hong Kong were British colonies. India (including what is now Pakistan) and Burma had a substantial measure of self-government, but their foreign affairs and many important decisions on internal matters were reserved for the British government in London. The whole of Indonesia was a Dutch colony, the Netherlands East Indies. France controlled Indo-China and small areas in India of which the chief were Chandernagore (above Calcutta) and Pondicherry. There were small Portuguese colonies in India (Goa, Damão and Diu), at Macao in China and in Timor. The colonial powers brought great benefits in the shape of stable, and for the most part honest, government, as well as improvements in communications, agriculture, irrigation, education and health. It is arguable that these areas would have done better to retain their dependent status. The ultimate responsibility for relieving poverty by economic development would have remained with the colonial powers, some of whom had already helped enormously and were well placed financially for further investment. But nationalist feeling is everywhere so strong that independence is preferable to whatever material advantages there were in the old system.

The end of colonial rule, always foreseen at any rate by Britain, was hastened by the war. Soon after 1945 India, Pakistan and Ceylon, followed later by Malaya and Singapore, achieved dominion status within the British Commonwealth. Burma became independent and chose not to remain in the Commonwealth. The Dutch yielded most of their ground to the new state of Indonesia. The French gave up all their land in India, and after a long war against local forces withdrew also from Indo-China. In 1961 Indian forces occupied Goa, Damão and Diu. The Philippines, which were dominated by Spain from 1565 to 1898 and thereafter by the U.S.A., were granted full independence in 1946. It remains to be seen whether the national status new to so much of the area will promote local pride and energy sufficiently to overcome the economic problems. Success so far has been uneven. India is well governed and has become a powerful international influence. At the other extreme, in Burma and Indonesia, central government is weak and large areas are often in armed revolt. Of the larger states which have never been controlled by western powers, China and Japan are powerful. Thailand is very backward but has an annual rice surplus, and poverty there is not acute by Asian standards. Finally, it is an unfortunate truth that nepotism and other forms of corruption are widespread at all levels throughout Monsoon Asia.

THE BOUNDARIES OF MONSOON ASIA

In the south and east, Monsoon Asia is bounded by the Indian and Pacific oceans. New Guinea belongs culturally to the Pacific realm, but the western half remained a colony of Holland until 1962 and was formerly part of the Netherlands East Indies. The Jakarta government hopes that the people of western New Guinea will decide at the referendum to be held by United Nations before 1969 to remain part of Indonesia. In western New Guinea the Pacific and Monsoon Asian realms overlap.

The northern boundary is a matter of opinion. It would be easy to map the inland limit of dense rural settlement in China and India. That would approximate to an east-west line a few miles north of Peking, turning southward about Lanchow to follow the eastern edge of the Tibetan plateau, ultimately to turn west again along the Himalayan foot-hills. All to the south of that line would be unquestionably Monsoon Asia. But much ground north and west of it is also best considered part of Monsoon Asia, for two reasons. First, in the Tarim and Dzungarian basins of High Asia there are oases, outliers of intensive agriculture with crops similar to those of lowland China. Second, a great deal of High Asia has been controlled by governments centred on the lowlands to the east and south. For centuries the Chinese have governed much of Mongolia and the country for 1,500 miles westward to the Pamirs, although few Chinese settled there. Between 1900 and 1947 the British Indian government was the most influential foreign power in Tibet. High Asia was dominated politically by China and Great Britain. Since 1950 Tibet has been occupied by the Chinese. The most realistic inland boundary for Monsoon Asia is the line to which Chinese authority extends, the political boundary of U.S.S.R. and the Mongolian People's Republic.

THE NAME 'MONSOON ASIA'

Recognizing a geographical region is one thing. Finding a good name for it is another. The word monsoon, probably derived from an Arabic word meaning season, was applied by the Dutch in the sixteenth century to the two principal seasons of northerly and southerly winds in the Indian Ocean. Today it also denotes the movements of air during those seasons. The monsoons influence practically the whole region as far inland as the Gobi Desert and southern Tibet. Unfortunately, however, the word monsoon is associated in popular literature with the hot, wet, inter-tropical countries only. This misleads school and university students into regarding Monsoon Asia as confined to the warmer rice-growing lands of the south, with disastrous consequences in examinations. Fundamentally it is bad practice to use a term from physical geography to describe a cultural entity. But Monsoon Asia is well established as a name. All the suggested alternatives such as Asia East by South, the Far East and the Orient should be avoided. They are either cumbersome or open to objection because they have been variously used to denote only part of the region. The name South-east Asia, in particular, should be reserved for the mainland and islands south of China and east of India.

2

Structure and Relief

There are three principal elements in the structural pattern of Monsoon Asia:

1—the older blocks;
2—the Alpine fold belt;
3—the tectonic depressions.

The older blocks are the visible parts of the foundations of the region. They are composed of plutonic, volcanic and metamorphic rocks, and hard sedimentary rocks. Some of the sediments were intensely folded and fractured at several stages far back in geological time. Today these areas are stable in the sense that they display no active volcanicity, are not subject to earthquakes and show no sign of recent dramatic upheaval.

In the Deccan of India, which is a stable block forming an immense plateau, granitic and metamorphic rocks predominate, and large areas are covered by thick horizontal beds of lava, the Deccan Traps. In Malaya, Thailand, Indo-China and Borneo, which are the higher parts of a stable block half submerged under shallow seas, granitic and sedimentary rocks are well represented.

The Alpine fold belt is built of sedimentary material, often with granites and schists in the cores of the folds. Many of the sedimentary rocks are relatively soft, partly because they are geologically young and, unlike those of the older blocks, have been subject to compression and folding during one orogenic period only. Yet some of the sedimentary rocks are as hard as any found elsewhere. The folding began in late Cretaceous times, reached its maximum in the Tertiary and continued in places into the Quaternary. These younger rocks, together with older, harder sedimentary rocks from greater depth that happened to get caught up in the earth movements, were folded, fractured, overthrust and raised to form mountain chains such as the Tien Shan, Karakoram, Himalayas and the mountains of Sumatra, Java, the Philippines and Japan. Minor earth tremors, causing slight damage to buildings, are frequent. Occasionally disastrous earthquakes occur, as at Tokyo in 1923 and at Quetta in 1935. Younger rocks are also found on the surfaces of the older blocks, but there they are undisturbed or at most only very slightly folded.

The tectonic depressions. Interspersed with the older blocks and the Alpine folds are great depressions. These have been partly filled with debris brought by rivers from higher ground. Enormous thicknesses of alluvium have accumulated. In places the alluvium extends beyond the limits of the depressions on to the fringes of the older blocks. In the Ganges delta, measurements of gravity anomalies indicate at least 6,500 ft. as the depth of the alluvium.

It is essential to remember that alluvium is not confined to these areas. It is found locally wherever the flow of streams is checked. Valley bottoms everywhere are likely to have some alluvium, but the large expanses measured in hundreds of miles are limited to the great depressions.

RELIEF

Figure 1 (page 25) illustrates the relationships between the three structural elements. It is intended for use alongside a good atlas relief map. The Tarim Basin and a few towns and mountain ranges are marked, to make comparison with other maps easier. One should not expect each of the three structural forms to be associated with one kind of relief only: the older blocks with plateaux, the Alpine folds with spectacular mountains and the tectonic depressions with alluvial lowlands. Each structural element, with the partial exception of the tectonic depressions, might have several kinds of relief, depending on the local circumstances of lithology and altitude in relation to surrounding areas.

The Alpine fold belt is a continuation of the Alpine fold zone of Europe. In the western half of the region it is characterized by plateaux and mountain chains. The Kirthar and Sulaiman ranges rise steeply from the Indus plains to 5,000 ft. The Hindu Kush in Afghanistan and the

Pamirs 100 miles north in U.S.S.R. reach 24,000 ft. Eastward from the Pamirs-Hindu Kush area four principal chains diverge. The Karakoram, its peaks including Mount Godwin Austen or K2 (28,250 ft.), strikes south-eastward into Tibet. To the south is the Himalayan system of several parallel ranges, with numerous peaks over 20,000 ft. and no passes lower than 15,000 ft. on to the Tibetan plateau. From Nanga Parbat (26,620 ft.) it runs first south-eastward and then eastward for 1,500 miles to end in Namcha Barwa (25,445 ft.), the mountain surrounded on three sides by the gorges of the Brahmaputra.

North of the Karakoram, the Altyn Tagh range overlooks the Tarim Basin. With the Nan Shan range farther east it forms a great arc of mountains on the northern edge of the Tibetan plateau. The fourth chain, the Tien Shan, with several peaks over 20,000 ft., forms the northern rim of the Tarim Basin and is associated with the narrow spur of Alpine folding shown on Figure 1. In addition the plateau of Tibet, lying between 11,000 ft. and 16,000 ft., has several mountain ranges, among them the Kun Lun, the Arka Tagh and the Kailas. The maximum width of the plateau between the Himalayas and the Altyn Tagh is 800 miles.

The trends of the great mountain chains in the Alpine fold belt are roughly the same as the prevailing strike of the folds and thrusts of their constituent rocks, that is to say at right angles to the direction of pressures from the adjacent older blocks which caused the folding. But, locally, powerful erosion has resulted in mountain ranges with alignments un-related to major structures. For example, in the eastern half of the Himalayas, rivers such as the Tista and the Arun have cut the country into north-south ranges over 15,000 ft. high and 50 miles long. More-over, it is in these circumstances that the relief is most impressive. The Tista, with its catchment in the rain-drenched Indian border protectorate of Sikkim, has carved a V-shaped valley over 13,000 ft. deep and 30 miles wide between the crests of adjacent ranges. In the steaming forests on the river bank at about 1,000 ft. one can part the branches of dense under-growth to see the ice slopes of the Kanchenjunga group rising to 28,000 ft. (or twice the height of the Matterhorn) only 40 miles away. All the vegetation zones, from tropical rain forest through coniferous timber to Alpine grassland and mountain tundra, are compressed into that short distance. The removal of vast quantities of rock to form many valleys of that magnitude has been accompanied by isostatic adjustment which has added considerably to what would otherwise have been the height of neighbouring masses such as Everest and Kanchenjunga. Most of the high peaks are of resistant igneous rock, but the upper slopes of Everest are of limestone which was indurated and compressed in the folding to become a calcareous schist.

Relief maps in atlases suggest that several ranges of Tibet continue

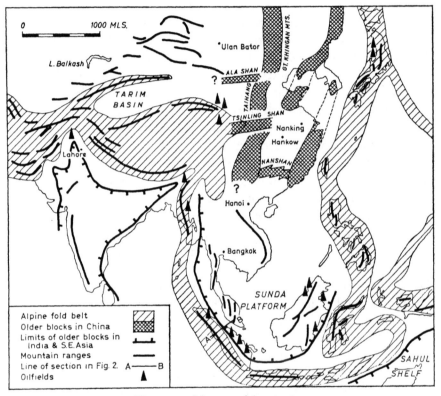

Figure 1. *Monsoon Asia: structure*

eastward into China while others bend south-eastward. With the exception of the western half of the Tsin-ling Shan of China, which is a range of young and old rocks folded in Tertiary times, they probably bear little relation to structure. Their alignments are thought to be the result of erosion by the headwaters of the Irrawaddy, Salween, Mekong and Yangtze which have cut the high ground of the area, partly of newer folded rocks and partly old blocks, into several elongated sections. The facts are uncertain because south-west Yunnan and eastern Tibet are re-mote and difficult country in which geological survey has made little progress. At any rate, the Alpine fold belt takes a sharp bend to continue through northern and western Burma where the multiple ranges of the Assam border are mostly about 6,000 ft. with peaks rising to 12,000 ft. Lower still are the Arakan Yoma farther south, with spurs running out from the main chain to end in sea cliffs which divide the coastal belt into isolated lowlands. These Burmese mountains are forested and very steep. Despite the lower altitude they have been more effective than the Himalayas as a barrier. Burma is more isolated from India than Tibet is. Until 1942 the border country was generally held to be impassable except for small parties travelling in the dry season. After the Japanese had occupied Rangoon in March of that year thousands of civilians died in attempts to walk out that way to India. It was not until General Orde Wingate, the Chindit leader, ordered that no area would be declared impenetrable until it had been penetrated that routes were established. Later those routes were improved temporarily to carry the motor transport of the 14th Army that recaptured Burma.

A striking feature of the Arakan coastlands are the mud volcanoes. Mud is extruded from vents; it is easily eroded, and active sites on the mainland are quickly reduced to concentric low rings of hard sediment, perhaps half a mile across. They are not dangerous. Cultivation extends right up to them. But, offshore, submarine vents are a hazard to shipping because the mud suddenly appears above the surface without warning. Unlike the mud volcanoes of Sicily and Iceland, which owe their existence to water heated at great depths, these Burmese features are not volcanic in the true sense. They are caused by gas escaping from oil deposits, and the mud originates very near the surface.

Beyond Burma other surface features are associated with the Alpine folds. Except for small areas in the Philippines there is hardly any plateau. Mountain ranges are submerged and form island arcs. Relief generally is strong, but there are flat coastal areas. Active volcanoes are numerous. On many islands much of the surface is covered with recent volcanic material, but there are also large areas of sedimentary, plutonic and meta-morphic rocks. The Arakan alignment can be traced through the steep ranges of the Andaman Islands, and on through the Nicobars to Sumatra where it appears as a double arc: an inner line forming the mountains of Sumatra; and an outer one through the numerous islands off the south-west coast, of which the largest are Simeuloee, Nias and Siberut. Other island arcs can be traced through Japan, Formosa and the 7,000 islands of the Philippines, and from Japan through the Bonins and Marianas to Halmahera and New Guinea. Where these Pacific arcs meet the Sumatra-Java alignment there are complicated structures which are reflected in the remarkable outlines of Celebes and Halmahera. If these localities of convergence were uplifted they would resemble the Pamirs and Hindu Kush in their relationship to mountain chains. The highest ground in Sumatra, Java and Japan is about 12,000 ft. and in the Philippines 9,000 ft. But the real relief of these submerged mountains is appreciated when one considers the depth of adjacent sea trenches. The Mindanao trough off the Philippines is 35,000 ft. below sea level, and similar depths occur off Japan. Depths of over 24,000 ft. occur about 200 miles off the south coast of Java. Seismic activity is widespread. About 1,500 minor shocks are recorded annually in Japan.

Evidence of very recent uplift is afforded by coral reefs. Live coral flourishes in clear shallow sea well away from the muddy water of river mouths. There are many instances of old reefs formed under such conditions now occupying elevated land. In Timor a reef of Quaternary age is at 4,265 ft. above sea level. Other areas have subsided on a similar scale to form deep seas.

In the island arcs there are hundreds of volcanoes, active and dormant. There are 17 active in Java, 10 in Sumatra, 12 in the Philippines and over 40 in Japan. Eruptions are mainly of ash; molten lava is not prominent. Death and destruction are sometimes brought to villages on the slopes by clouds of hot gases and avalanches of mud. Really spectacular eruptions are rare. The last was in 1883 when Krakatau, an island five miles across in the Sunda Strait, blew up with a bang that was heard 1,800 miles away in Ceylon. Ash fell over southern Sumatra and as far away as Singapore. Great sea waves flooded neighbouring coasts to a depth of 100 ft. Over 36,000 lives were lost. Lesser disasters are more frequent. In 1930 hot gas and mud from the 8,000 ft. cone of Merapi in Java destroyed a score of villages. There are observation posts on many crater rims from which warning by phone is given when the volcano becomes unusually active or heavy rain threatens to cause dangerous mud flows.

The area of older blocks north of the Alpine folds includes a vast expanse of plateau in the Tarim Basin and Mongolia, the geological detail of which has not been worked out. Most of the plateau is between

3,000 ft. and 5,000 ft., but the bottom of the Turfan depression in Sin-kiang is 928 ft. below sea level. On the northern borders there are several mountain ranges including the Altai and Sayan.

Farther east Chinese geologists recognize the older blocks shown in Figure 1. These blocks are of two kinds: the ones aligned north to south or north-east to south-west, and those running east to west. The first group consists of three geanticlines which have been stripped by erosion to reveal very old hard rocks along their axes. For the most part they form mountainous country up to 5,000 ft., with strong relief, especially where they are near the sea, as in the highlands of south-east China and in the Shantung peninsula. Inland, the westernmost geanticline can be traced from the province of Yunnan northward through the Taihang range to the Great Khingan mountains that overlook the lowland of Manchuria. A central geanticline forming the Shantung peninsula has a submerged portion under the Gulf of Chihli, but it reappears in the Liaotung mountains of eastern Manchuria. The easternmost geanticline can be traced into Korea. There are three upfolded areas of east-west alignment: these form the Ala Shan, Tsinling Shan and the Nan Shan of south China, the last not to be confused with the Alpine Nan Shan farther north.

South of the Nan Shan block is the great stable block around which the island arcs were folded. Its limits are well defined in Burma where the edge of the Shan plateau is a fault scarp facing the Tertiary lowlands of the Sittang and Irrawaddy. The block forms mountainous country in the Annamite highlands of Indo-China and low plateau in the Korat area of Thailand. The southern part of the block, including Malaya and Borneo which are both mountainous, is called the Sunda Platform. Most of this platform lies below the shallow South China Sea.

South of the Alpine folds there are two areas of older block country: peninsular India and Ceylon, and northern Australia. Peninsular India is mostly plateau between 2,000 ft. and 4,000 ft., tilted downwards to the east so that the western edge stands out as a 5,000 ft. range of mountains, the Western Ghats. The descent to the west coast is steep, with deep gorges and waterfalls. The block extends under the Bengal plain to re-appear in the Shillong plateau, while in the north-west under the Indus plain it forms a foundation round which the Alpine folds were raised. Northern Australia is part of another block, the Sahul Shelf, the southern counterpart of the Sunda Platform.

The tectonic depressions are occupied either by sea or alluvial low-lands. They stand out well on atlas relief maps. The lowlands are extend-ing at the expense of the sea, as a result of continued deposition by the rivers. They are of outstanding significance in the geography of Monsoon Asia because they of all the major land forms offer the best opportunities for agriculture. Most of them support dense rural populations. Two great lowlands dwarf all the others: the Indo-Gangetic plains and the North China plain. Their size can be appreciated by comparing their dimensions with a familiar distance elsewhere, say the 400 miles from Brighton to Berwick-on-Tweed. From the Himalayan foothills south-west along the Indus plain to the delta is 700 miles. From the foothills of North-West Frontier Province south-eastward through the Ganges lowland to the hills of East Pakistan is 1,400 miles. These plains are about 200 miles wide. The lowest point on the Indus-Ganges divide is little more than 700 ft. The average slope to the sea along the Indus is less than one foot to the mile, and about half that figure along the Ganges. The surface of the North China plain is broken by the Shantung peninsula, but there is continuous plain for 500 miles from the southern end of the Taihang range to the sea, and for 800 miles from Peking southward. The absence of solid rock at the surface over these great distances is a handicap in building roads. Fortunately the river silts make quite good bricks.

Among the smaller lowlands are those of the Red river and the Mekong in Indo-China, the Menam Chao Praya in Thailand and the Irrawaddy and Sittang in Burma. All these are lowlands associated with one major river. There are also some large ones constructed jointly by several smaller rivers. For example, the alluvial lowland of eastern Sumatra, over 800 miles long and in places 100 miles wide, is the result of deposition by several relatively short rivers. The lowlands of southern New Guinea are of similar origin.

MINERAL RESOURCES

The detail of mineral resources is best considered later when individual countries are discussed. There is the striking fact, however, that each of several minerals is associated with one major structural unit only. All the known oilfields, as shown in Figure 1, are related to the Alpine fold belt. They occur not in the centre of the belt but at the margins and on adjacent parts of the older blocks. The explanation is as follows. Apart from small accumulations in Carboniferous rocks in Szechwan, oil is usually found in the younger rocks. But the Tertiary rocks in the centre of the fold belt are so faulted, fractured and overthrust that no unbroken anticlines exist. However, the Tertiary seas extended over neighbouring parts of the older blocks, which thus acquired in places a layer of newer sediments. During the mountain-building process the edges of the older blocks formed stable foundations which protected the newer sediments above them from the more extreme effects of compression and folding. Instead of being intensely disturbed, the newer rocks were buckled into gentle folds free of faults and thrust planes: and where impermeable rocks are interbedded with porous strata these structures form good oil reservoirs. The geological section in Figure 2 illustrates the relationship of oilfields to underlying structures in Sumatra. Similar explanations apply to the oilfields of West Pakistan, Assam, Burma, Sinkiang (associated with the Tien Shan), Yumen in China (near the Altyn Tagh-Nan Shan range) and in Java, eastern Borneo and Japan. The only oilfields that do not fit clearly into this picture are those of British Borneo, which appear on Figure 1 to be 400 miles from the Alpine belt. These fields are probably related to subsidiary Tertiary fold axes which link up through Palawan with the Philippines.

Of the Alpine fold areas, only Japan has useful quantities of coal, and even there the seams are thin, faulted and difficult to work.

Of the older block areas, peninsular India is exceptionally rich in iron ore of the highest quality, and has good coal. The iron ore is in the very old rocks, and the coal is in terrestrial deposits (the Gondwana rocks) of Carboniferous, Permian and Triassic age which have survived erosion in down-faulted areas principally in the north-east. Manganese is plentiful. Ceylon, which has structural affiliations with India, is unfortunate in having no coal and hardly any iron ore. In China coal and iron ore are widespread, but the chief reserves are in the north and in Manchuria. The Sunda Platform and its northern upland extensions into the Shan States of Burma and into Yunnan are remarkable in that they have no good coal except in North Vietnam, but are rich in tin ore. This ore occurs in veins in granite masses and the surrounding country rock in Yunnan and in much of the platform country south of Rangoon. The veins are difficult to locate and work, especially in the humid, forested south where they are concealed by deep layers of weathered rock. However. in these southern

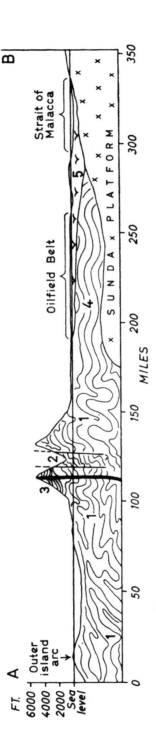

Figure 2. *Geological section: Sumatra* I—Tertiary and older sedimentary rocks, intensely folded and fractured, in the central part of the Alpine belt; 2—the longitudinal valley of the Sumatra highlands; 3—volcanic ash and lava of the western range, with a vent leading down to the magma; 4—Tertiary sedimentary rocks gently folded on a stable foundation of Sunda Platform; 5—alluvium.

This is a simplified section along line A–B of Figure I, showing the position of the oilfield belt of gently folded Tertiary rocks. Oil accumulated about the anticlinal axes. In Sumatra, where these structures are partly masked by great thicknesses of alluvium and mangrove mud, groups of oil wells are a characteristic sight in the mangrove and swamp forest of the eastern lowland (see Figure 48).

areas the veins have been eroded along with surrounding rocks and the ore has been conveniently sorted and deposited as alluvial accumulations by streams in their lower courses. Most of the tin ore produced in Malaya, Thailand and Burma and in the Indonesian islands of Banka and Billiton, which are also part of the Sunda Platform, is from alluvial workings.

This matter of the relationship between mineral resources and structure is significant in the economic geography of Monsoon Asia. Clearly, states such as China, Burma and Indonesia, each of which embraces parts of more than one major structural division, are likely to be better off as regards the mineral bases for industrialization than Malaya, Thailand or Ceylon, whose territories are each confined to one of the divisions.

3

Climate

THE MONSOONS

In winter atmospheric pressure is high over central Asia and Indo-Pakistan, and low over the oceans to the east and south. The movement of air from the high to the low pressure areas gives westerly winds over north China and Japan, and north to north-east winds over India, the Bay of Bengal, the Philippines and parts of Indonesia. These winds are the winter monsoon. In summer low pressure exists over the continent and high pressure over the oceans. Air movement from the high towards the low pressure in that period gives southerly winds—south-west off the west coast of India, south in the Bay of Bengal, and south to south-east over China, Japan, the Philippines and Indonesia. These winds are the summer or southerly monsoon. The words winter and summer might seem inappropriate labels for the monsoons in the equatorial part of the region where there are no substantial changes in mean temperatures during the year. Alternatives such as north-east and south-west, however, are less acceptable, for although they apply well enough over the Indian Ocean they are often meaningless for the land areas, because both monsoons vary greatly in direction from place to place.

RAIN

In Monsoon Asia rain occurs mainly in association with six atmospheric phenomena:

1—the converging airstreams of the equatorial zone;
2—the summer monsoon;

33

3—the winter monsoon;
4—winter depressions;
5—typhoons;
6—hot-weather depressions.

The converging airstreams of the equatorial zone, both of warm, moist air in the form of the north-east and south-east trades, bring convectional and orographic rain to areas within eight degrees of the equator. Both kinds of rain occur as heavy downpours. The high temperatures promote strong convection currents, seen in intense form as cumulo–nimbus cloud towering to 20,000 ft. and more. Vertical air currents, reinforced by heat released by condensation of water vapour into raindrops, sometimes exceed 100 miles an hour and are a hazard to air navigation. There have been several instances of aircraft breaking up when flying unavoidably through such cloud. In southern Malaya and western Java the day usually begins with bright sunshine. By 10 a.m. convection results in partial cloud cover which does not disperse until evening. Total hours of sunshine are high for so wet an area. Jakarta gets on an average 2,326 hours a year, over double the figure for parts of west Scotland. Violent thunderstorms are frequent. The average in a year is 130 at Jakarta and 220 at the hill station of Bandoeng, 80 miles away at 2,400 ft.

About 10 years ago a theory became popular that this equatorial rain was connected with an inter-tropical front between the converging trade winds and analogous to the polar front of higher latitudes. The inter-tropical front was considered to move north and south with the annual migration of the sun. However, the concept proved to be of little use in forecasting weather, and doubts arose as to the value of applying meteorological concepts of high latitudes to the very different conditions at the equator, especially as the airstreams involved have such similar physical properties that the front is hard to locate.

The summer monsoon. The low pressure over northern India is well established by May, but high pressure lingers over the Indian Ocean between the Malabar coast and Africa. Progressive heat from the sun on its way to the northern tropic reduces that high pressure to a stage at which an uninterrupted pressure gradient prevails from south of the equator to the centre of the low over northern India. When that condition is achieved the summer monsoon begins. At Colombo in 1938 weak and indeterminate winds prevailed until 29th April when the summer monsoon arrived as a south-west wind of up to 25 miles an hour average velocity which persisted into October (Figure 3). Maximum wind force is such that in the exposed artificial harbour at Colombo mooring buoys are carefully arranged so that ships can lie facing into wind. Liners tied up across wind drag their moorings and damage other vessels.

Northward, the date of the onset of the monsoon is progressively later. At Bombay it is usually about 5th June. Meanwhile, by mid-June a second branch of the monsoon in the Bay of Bengal reaches Calcutta. Some of this airstream crosses the Himalayas to give rain in south-east Tibet where Lhasa sometimes gets 60 in. a year, but most of it is deflected along the Ganges valley towards the low pressure of Sind, reaching Delhi before July. Along the southern edge of the Ganges plains, where the first branch moving inland from the west coast impinges on this second branch, rainfall is increased locally by frequent storms.

In south China the monsoon begins in May but there is no sudden onset. Also there is less rain, partly because the winds have already lost moisture over Indonesia, Indo-China and the Philippines. By the time the monsoon reaches Peking it has little moisture left. Two hundred miles farther north is the Gobi Desert.

By late September or early October the summer monsoon rain has ended in north China, Burma and northern India. Farther south the summer monsoon is still recognizable. Disturbances associated with its retreating northern edge give useful rain in Madras in October and November.

The summer monsoon is a dominant feature bringing rain to all areas in its path. For nearly half the year it replaces the system of converging air masses at the equator.

The winter monsoon brings dry polar air into north China. Some of the cold air penetrates occasionally well south of the Tsinling Shan. In Indo-Pakistan, screened by the Himalayas from the cold air of central Asia, the winter monsoon consists of local air and is warmer. The system is well established over China, Japan, the Philippines and most of India by mid October; in November it also embraces Ceylon and Malaya, and by January it replaces the converging trade winds over Java. In the record for 1938 at Colombo (Figure 3) it was established as a constant and powerful north-east wind by the end of November and continued into February.

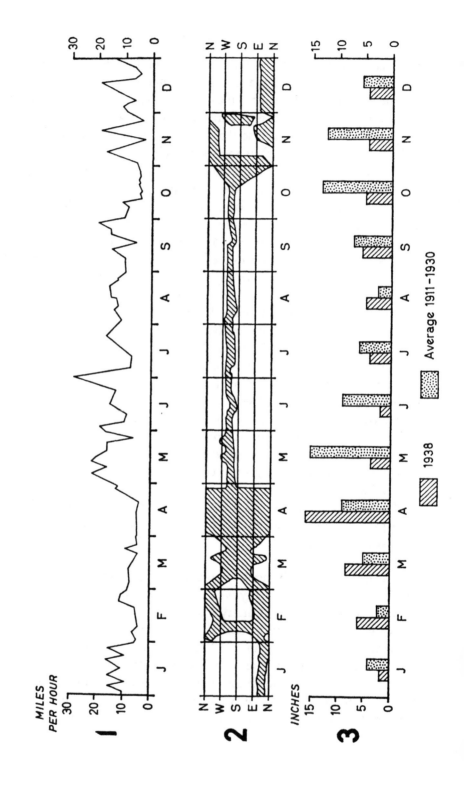

MILES
PER HOUR

1

INCHES

2

3

Average 1911–1930

1938

Figure 3. *Colombo: climatic data for 1938* 1—wind velocity; 2—wind direction; 3—monthly rainfall.

Graphs 1 and 2 refer to the Pilot Station on the southwest breakwater of Colombo harbour where the vane of the anemometer was 95 ft. above the water, and 3 refers to the Observatory in the town.

The graph of wind velocity shows the *mean rate* for each day. *Maximum* velocities were 65 m.p.h. on 20th February, and 56 m.p.h. on 30th June and 4th July.

The second graph is based on records of all changes in wind direction during the year. A full-width band of shading indicates changeable winds from all quarters. A narrow band of shading indicates wind of constant direction. Notice how indeterminate winds of February, March and April coincide with low values of mean wind velocity. The summer monsoon is prominent as a narrow band of shading between the lines S and W (indicating constant SW wind) and coincides with high wind velocities. The winter monsoon is well marked on graphs 1 and 2 in December and January.

On the third graph, the rainfall of 1938 compares unfavourably in most months with the averages for 1911–30. The average annual rainfall at Colombo is 90·8 inches. The total for 1938 was only 64·76 inches.

The onset of the winter monsoon at Colombo is the signal for the harbour authorities to rearrange their moorings to avoid the north-east wind.

The winter monsoon is associated with dry weather, except in Japan, the Philippines, southern China, Indo-China, Malaya, Indonesia and Ceylon and the southern tip of India, to all of which it brings rain after gathering moisture from the surrounding seas. Wind speeds are lower than during the summer monsoon, except over China and the China Seas, where gales are frequent.

The winter depressions enter West Pakistan from Afghanistan and the Persian Gulf, and follow a well-defined track across the Punjab. Many fade out over the Ganges plains but some reach the Brahmaputra. Rain associated with them rarely amounts to more than two to three inches on the plains, but even that is useful to farmers at the cool season when evaporation losses are low. The hills to the north get as much as 10 in. Similar depressions, probably originating on the border of Tibet, move across China and Japan. The most frequented track is along the Yangtze valley. The Chinese depressions give more rain as they move eastward and attract air from the China Seas.

Typhoons, or tropical cyclones, originate in the western Pacific, move westward over the Philippines and approach the south-east coast of China on parabolic courses. Some recurve northward and reach Japan. Winds attain terrific force and do great damage. About 100 typhoons are recorded each year, mostly between June and November. Rainfall associated with them is experienced over a much wider area than the high winds, and up to 200 miles inland in southern China.

Similar storms, referred to locally as cyclones, occur in the Bay of Bengal. About a dozen each year, they move north, chiefly in July, August, September and October. Occasionally tremendous seas have overwhelmed coastal settlements and crops have been destroyed in the fields. In 1876 100,000 people died within half an hour. In October 1960 over 10,000 were lost in two cyclones which swept over Chittagong, Noakhali and Barisal.

Hot-weather depressions. The hot weather is the period immediately before the summer monsoon, when the northbound sun in the cloudless skies of March, April and May brings intense local heating and convection. Violent duststorms with no rain develop in the Punjab and the

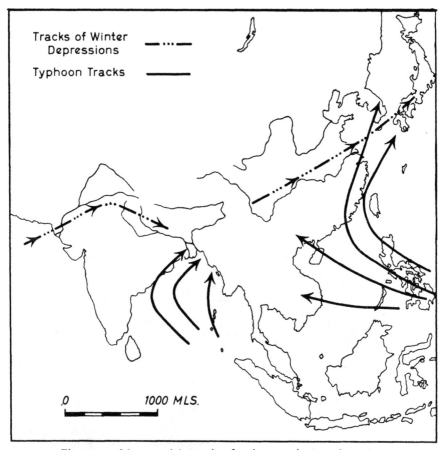

Figure 4. *Monsoon Asia: tracks of typhoons and winter depressions*

Ganges lowland. Storms of similar mechanism in the Ganges delta and Assam attract moist air from the Bay of Bengal and give substantial rainfall, agriculturally of great value, before the main wet season begins. In Burma duststorms are frequent. In south China and Japan depressions of local origin give rain at this season.

RAINFALL DISTRIBUTION

The typhoon and winter depression tracks are represented in Figure 4, while Figure 5 shows the average annual rainfall of Monsoon Asia. High rainfall is characteristic of the equatorial areas, and of coastlands and high ground exposed to the summer monsoon. In general rainfall decreases towards the interior of Asia. But diagrams of that kind are of little value

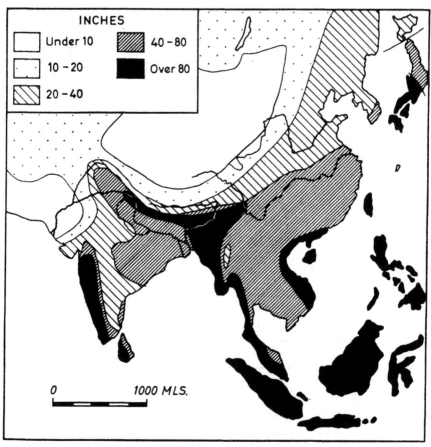

Figure 5.   *Monsoon Asia: average annual rainfall*

on their own. It is essential to look at the rainfall records of as many places as possible. To that end the following paragraphs deal in turn with each of the 28 stations in Table 1. The location of each station in relation to major relief features should be noted from atlas maps.

Singapore has rain in every month and an annual average in keeping with its equatorial location. There are three maxima, in May, August and December. The first two maxima, related to the annual migration of the

MONSOON ASIA: AVERAGE RAINFALL (inches)

	Lat.	Alt. (ft.)	J	F	M	A	M	J	J	A	S	O	N	D	Year
Singapore	1°N	10	9.9	6.9	7.6	7.4	6.8	6.8	6.7	7.7	7.0	8.2	10.0	10.1	95.1
Kota Bharu	6°N	10	10.2	5.5	7.0	4.6	6.4	6.1	5.5	6.6	8.7	12.0	24.0	26.3	122.7
Alor Star	6°N	8	1.7	1.2	5.3	10.5	8.6	7.5	6.6	10.6	10.2	13.7	8.3	6.2	90.4
Penang	5°N	23	3.5	3.0	5.8	8.2	10.3	7.4	6.6	10.8	14.1	16.6	12.0	5.9	104.2
Mergui	12°N	24	1.1	1.9	2.3	5.5	15.2	30.0	32.7	29.1	24.0	11.6	4.3	1.3	159.0
Rangoon	17°N	18	0.2	0.2	0.3	1.4	12.1	18.4	21.5	19.7	15.4	7.3	2.8	0.3	99.6
Quang Tri	17°N	23	6.0	1.2	1.5	1.3	2.8	2.6	4.8	3.9	15.7	26.4	16.0	7.2	89.4
Akyab	20°N	20	0.0	0.1	0.3	0.8	19.0	37.9	47.4	37.0	26.3	11.5	3.0	0.9	184.2
Mandalay	22°N	250	0.1	0.1	0.2	1.1	5.8	5.5	3.3	4.6	5.7	4.7	1.6	0.4	35.1
Calcutta	23°N	21	0.4	1.0	1.4	2.2	5.6	11.9	12.7	13.4	10.0	4.9	0.6	0.2	64.3
Cherrapunji	25°N	4,309	0.7	2.3	10.6	31.3	50.8	103.6	107.4	81.5	49.4	16.8	2.3	0.3	457.0
Darjeeling	27°N	7,376	0.6	1.1	1.8	3.8	8.7	24.9	32.3	26.1	18.4	4.5	0.8	0.2	122.7
Dacca	24°N	20	0.3	1.2	2.4	5.4	9.6	12.4	13.0	13.3	9.8	5.3	1.0	0.3	74.0
Allahabad	25°N	300	0.7	0.5	0.4	0.1	0.3	4.7	12.0	11.0	6.3	2.3	0.3	0.2	38.8
Lahore	32°N	702	0.9	1.0	0.8	0.5	0.7	1.4	5.1	4.7	2.3	0.3	0.1	0.4	18.1
Simla	31°N	7,232	3.6	3.7	3.3	2.7	3.9	8.8	21.1	20.7	7.5	1.4	0.5	1.3	79.3
Colombo	7°N	24	4.0	2.2	4.9	8.9	15.0	9.0	6.0	2.6	7.0	13.2	12.3	5.6	90.8
Watawala	7°N	3,259	5.6	2.9	7.3	11.0	21.9	40.7	29.8	26.7	27.0	24.0	15.1	7.3	219.5
Bombay	19°N	37	0.1	0.1	0.0	0.0	0.7	19.9	24.2	14.5	10.6	1.9	0.4	0.0	72.4
Karachi	25°N	13	0.6	0.4	0.3	0.1	0.1	0.6	2.8	1.7	0.6	0.0	0.1	0.2	7.5
Madras	13°N	22	1.1	0.3	0.3	0.6	1.8	2.0	3.8	4.5	4.9	11.2	13.6	5.4	49.5
Bangkok	14°N	22	0.1	0.8	1.4	3.1	4.7	6.4	5.9	5.6	11.1	7.5	2.2	0.5	49.3
Chiang Mai	19°N	2,400	0.4	0.1	1.0	1.6	6.0	5.1	5.7	9.1	9.5	6.7	2.3	0.5	48.0
Foochow	26°N	66	1.7	3.6	4.6	4.9	5.9	7.6	6.5	7.8	8.0	1.8	1.6	1.9	55.9
Ichang	31°N	371	0.8	1.2	2.1	4.0	5.0	6.3	8.2	7.0	4.2	3.1	1.4	0.7	44.0
Peking	37°N	125	0.1	0.2	0.3	0.6	1.3	3.3	9.8	5.7	2.3	0.7	0.3	0.1	24.7
Tokyo	36°N	70	2.0	2.6	4.3	5.3	5.9	6.3	5.6	4.6	7.5	7.2	4.3	2.3	57.9
Surabaya	7°S	16	12.3	11.5	10.5	7.4	4.3	3.5	1.9	0.5	0.5	1.5	4.6	9.8	68.3

Note: Monthly figures of two inches and less are underlined to show approximate duration of dry seasons

sun, are only slightly above the figures immediately before and after. The third and most prominent, of 10·1 in. in December, is the result of the replacement of the normal equatorial system of converging trade winds by the winter monsoon of moisture-laden northerly winds from the Pacific Ocean and the China Seas. The summer monsoon becomes established near the equator early in May but is not marked by a rainfall maximum at Singapore. At that southern fringe of the summer monsoon system winds are only light southerly breezes of local air, not highly charged with moisture after a long sea crossing: and at that time Singapore is in the rain shadow of the Sumatra highlands.

At *Kota Bharu* on the north-east coast of Malaya the contribution of rain from the monsoons and the converging airstreams results in three maxima. The high maximum of 26 in. in December is brought by the winter monsoon. The whole of Malaya is in the rain shadow of Sumatra during the summer monsoon. The maximum early in the year is contributed mainly by the greater convection associated with the northward passage of the sun in March and April. That explains the maximum of seven inches in March at Kota Bharu, but on the west coast enough rain is brought by the south-west winds of the summer monsoon in late April and May to move the early maximum to April at *Alor Star* (10·5 in.) and May at *Penang* (10·3 in.). The additional rain-shadow effect of Malaya itself in relation to the summer monsoon is illustrated by the comparatively slight maximum of 6·4 in. at Kota Bharu in May.

Reviewing the first four stations together, the winter monsoon emerges as the strongest single influence on the trend of the rainfall figures in the equatorial zone. An exceptional feature is the low rainfall in January and February at Alor Star where those months are much drier than at Penang, only one degree farther south. Six degrees north of Alor Star, at *Mergui* on the Tenasserim coast of Burma, there is a single maximum in July; and despite the high aggregate of 159 in. the period of the winter monsoon is remarkably dry with three months each with less than two inches. At *Rangoon*, five degrees farther north again, the pattern is the same, but there are five consecutive months with less than two inches of rain. Northward from Penang the winter monsoon becomes a dry season of increasing duration, and the annual rainfall becomes concentrated into one wet season associated with the summer monsoon. *Quang Tri* in Indo-China, however, at the same latitude as Rangoon, is exposed to the winter monsoon blowing across seas, but is sheltered from the summer monsoon. There the winter monsoon brings most of the rain, and a subsidiary maximum in July is caused by the summer monsoon which arrives from the south-west.

The next two stations, *Akyab* and *Mandalay*, are of special interest as an example of rain shadow. The figures indicate dry seasons of similar length and intensity. While the strong relief of the Arakan mountains induces a tremendous rainfall from the summer monsoon over the little port of Akyab, that same airstream is so dry on reaching the Irrawaddy as to give only one sixth of the Akyab figure at Mandalay.

Yet another interesting aspect of the summer monsoon is illustrated by the next three stations: *Calcutta, Cherrapunji* and *Darjeeling.* Calcutta, with an aggregate of 64 in., is representative of places on the Ganges delta well away from hills lying across the path of the monsoon. Cherrapunji, an exposed station at 4,300 ft. on the Assam plateau, gets the enormous annual average of 457 in. with a July maximum of 107 in. Over 900 in. have been recorded in exceptional years. Yet at Darjeeling, equally exposed but 3,000 ft. higher, the total is only 122·7 in. The explanation is that the monsoon airstreams are shallow. They give maximum rainfall within 5,000 ft. of sea level. Above that altitude monsoon rain gets progressively less, and precipitation on the higher slopes of the Himalayas is not heavy.

Now compare those three stations with Rangoon, Akyab and Mandalay. The wet season of the Burmese stations is seen to begin suddenly, with a sharp rise in May from less than two inches in April indicating the arrival of the monsoon. At the Indian stations, where the monsoon breaks in June, there is a gradual increase in rainfall through March and April into May. In those three months Calcutta, for example, receives over nine inches and *Dacca* 17 in. That is the rain from the hot-weather depressions.

Allahabad, Lahore and *Simla* are in the path of the monsoon airstream deflected along the Ganges lowland towards the low pressure of Sind. On the plains the monsoon rain decreases north-westward. At Allahabad the aggregate is below 40 in. and at Lahore less than 20 in. Both stations have slight winter rainfall making a subsidiary maximum in January and February; that is the rain from the winter depressions. It is greater in the Himalayan foothills, amounting to about eight inches at Simla (7,200 ft.), where the mountains also induce a larger contribution from the monsoon.

Rain occurs throughout the year at *Colombo* and *Watawala* in Ceylon.

It is brought by both monsoons. In February, March and April, and also in October, when neither monsoon is operative, it is associated with the slacker winds of the converging trades. There is a double maximum at Colombo: one in May associated with the burst of the summer monsoon, the other in October-November just before the onset of the winter monsoon. At Watawala in the hills the summer monsoon rain brought about by the local relief is so great and persistent that the figures show a single maximum in June of 40·7 in.

A thousand miles farther north, at *Bombay*, the winter monsoon is virtually a rainless period. *Karachi* gets a low aggregate of 7·5 in., partly because summer monsoon air reaching that area arrives from the north-east after moving counter-clockwise round the low pressure of Sind and losing much moisture in the process. What moisture is left is dissipated by hot dry upper air from the Baluchistan plateau which flows south over the shallow monsoon system. The January figure of 0·6 in. is related to the winter depressions.

At *Madras* the dry season is long, and the single maximum in November is related to the retreating front of the summer monsoon.

Bangkok and *Chiang Mai* are remarkable for their low aggregates in comparison with Rangoon. Much of Thailand is in rain shadow during both monsoons.

Of the three Chinese stations, *Foochow* is interesting for its three maxima and a rainfall well distributed over the year. Most of the rain is from the summer monsoon with a maximum in June. Typhoons partly account for the September maximum, and the minor peak in December is related to the winter depressions. At *Ichang* in central China there is summer monsoon rain, and depressions moving eastward along the Yangtze give a modest winter rainfall. *Peking* is entirely dependent on a depleted summer monsoon. Eight months of the year are dry.

At *Tokyo* there is rain throughout the year contributed by the two monsoons, the winter depressions from the mainland and, in September and October, the typhoons. Stations on the west coast of Japan have higher figures for the winter monsoon period.

The last station, *Surabaya* in the east of Java, is far enough south to come under the influence of dry southerly winds from Australia between June and October. Eastern Java and the islands as far as Timor have a dry season at that period.

In the equatorial zone mean annual temperatures are around 80°F.: 81°F. at Singapore, 79°F. at Pontianak and 80°F. at Colombo. Mean annual ranges are slight: Singapore 2·3°F.; Pontianak 2·2°F.; Colombo 3·6°F. Shade temperatures are never very high. The absolute maximum recorded at Singapore is 94°F., and the absolute minimum 67°F., but the daily range is usually about 11°F. Humidity is high, and for Europeans the lowland climate is uncomfortable except at the coasts where local sea breezes bring relief during the day. After sundown land breezes prevail and the nights are oppressive. Life is more tolerable in the hills. Near Cameron Highlands at 4,750 ft. in Malaya the mean annual temperature is 64°F., night temperatures are about 50°F. and 36°F. has been recorded. Similar conditions are found in Ceylon, where at 4,000 ft. the evenings are cool enough for fires to be a welcome sight in planters' bungalows.

Northward as far as Calcutta mean annual temperatures are much the same: Rangoon 81°F., Mandalay 82°F., Bombay 81°F., Delhi 77°F. and Calcutta 71°F. The annual range is greater: 10°F. at all except Delhi where it is 34°F. The highest temperatures occur in the three months before the summer monsoon. Daily maxima above 90°F. are common. In the Indus lowland 120°F. is often exceeded. The diurnal range is about 15°F. at Bombay, rising to 30°F. at Delhi. On account of the cloud cover, maximum temperatures are slightly lower during the summer monsoon but humidity is higher and it is a matter of opinion whether the wet-season conditions are the more bearable. In winter clear skies permit high daily maxima, often reaching 80°F. Radiation at night is rapid. Frosts occur between December and February in north India and the Shan plateau of Burma.

In China the mean annual range is greater because of the cold winter monsoon. The severity of winter in the north can be judged from the Peking figures: the averages for December, January and February are below freezing, an absolute minimum of −5°F. has been recorded, yet the July average is 80°F. The winter monsoon carries the cold weather far south, although the Tsinling Shan affords some protection to the Yangtze valley. The July averages for Ichang and Lahore, two towns at nearly the same latitude, are 84°F. and 89°F. respectively, but the corresponding figures for January are 45°F. and 53°F. Slight frost has been recorded as far south as Hong Kong and Foochow.

The inner Asian borders of Monsoon Asia have extreme conditions.

At Ulan Bator in Mongolia mean temperatures are 63°F. for July, −2·6°F. for January, and they remain below freezing for six consecutive months. A temperature of −30°F. has been recorded. The oasis of Kashgar is remarkably mild: the average for July is 80°F.; and for December, the only month below freezing, 25·6°F. In Tibet monthly averages rarely exceed 60°F. and they stay below freezing for two to three months.

RAINFALL RELIABILITY

The rainfall of a district in a single year is likely to differ from the average. In Monsoon Asia the differences are often large. Generally, the lower the average rainfall the greater the variability, but in agricultural areas of very low rainfall irrigation is usually well developed and drought is therefore not a serious menace. Over much of India and China, where the normal rainfall of between 30 in. and 50 in. is just adequate for grain crops and there is less irrigation, a deficit of 50 per cent can bring disaster. The peasants might have to sell draught animals and their few other possessions to buy imported food. Modern communications and international relief now prevent the worst effects of famine, but only 40 years ago deaths by the thousand were a commonplace. A delay in the start of a wet season can be as harmful, even if the aggregate for the year is normal. If rain greatly exceeds the average, destructive floods occur. Hardship is greatest in peninsular India and north China, both dependent almost entirely on the summer monsoon. Where there is rain from more than one source, as in south China where winter depressions, typhoons and the monsoons give moisture, all the sources are unlikely to fail in the same year. Even in the equatorial zone large variations are experienced. Figure 3 shows the mean monthly rainfall at Colombo and the amounts received each month in 1938 when the total was only two thirds the average. In February, March and April of 1938 rainfall was well in excess of the average, but the first three months of the summer monsoon, and also October and November, were abnormally dry. In these areas variations are not so disastrous because there is usually enough rain for crops to grow. Yields are affected. In Malaya the exceptionally dry February and March of 1959 brought an immediate decline in the output of local rubber trees, and this was reflected in a substantial increase in rubber prices at the Singapore markets.

4

Natural Vegetation

On the northern fringe of Monsoon Asia is a wide belt of desert and semi-desert reaching from the western end of the Tarim Basin for 2,500 miles through the Taklamakan and the Gobi to the western border of Manchuria. There is also desert in the Indus valley of West Pakistan and neighbouring parts of India. In the country bordering the deserts rainfall is low and the natural vegetation is grass. In the grassland and desert of the Tarim Basin there are oases with natural woodland wherever streams from nearby mountains bring additional water. In the Himalayas above 13,000 ft. and in Tibet there is mountain grassland and mountain tundra vegetation. The plant cover of these northern areas, together with the local relief, was an important element in historical geography. The oases determined the alignment of the land routes from China to the west, the mountains and higher plateaux were barriers to movements of people, and the grassland early became the home of nomadic people whose periodic invasions of the richer agricultural lands to the south had profound effects in India and China. Over the remaining, vastly larger part of Monsoon Asia the natural vegetation is forest. In crowded areas, especially China, India, Pakistan and Java, the forest has been completely cleared from enormous areas to make room for cultivation. Other countries, notably Malaya, Sumatra, Borneo and Burma, are still well wooded.

Tropical rain forest, evergreen and enormously rich in species, but without relatively pure stands of one kind of timber and consequently not of great economic value, occurs in the equatorial areas that have heavy rain throughout the year. The taller trees reach 200 ft. in height. Lianas cling to the trunks and trail from the branches, and epiphytes abound. Where there are plenty of trees to cut off most of the direct sunlight there is little undergrowth. In clearings for roads and railways strong sunlight reaches the ground and promotes a dense growth of bushes up to 40 ft. high which persists for only a few hundred yards into the tall timber. In very wet areas farther north, as along the Western Ghats, the Arakan Yoma and the foothills of the eastern Himalayas, the forest is similar in appearance to the equatorial form despite the dry season.

Except in uninhabited areas, such as central Malaya, very little of the rain forest seen today is natural in the sense that it has reached an ecological climax without human interference. There are usually some shifting cultivators who clear timber from small plots, raise crops there for a year or so until the soil is exhausted and then move on to clear and plant fresh ground. The abandoned plots first become colonized by grasses and bushes that flourish in strong light. At a later stage trees appear. Left to themselves the plots would revert to a climax rain forest in about a century, but often they are cleared again after 20 years or so. Occasional destructive fires after short spells of dry weather have similar results. In Borneo and northern Sumatra there are hundreds of square miles of coarse *alang* grass where the annual rainfall is great enough for forest. Those areas were once forested, but after repeated clearing, burning and cultivation the soil became too poor for trees to re-establish themselves.

Tropical monsoon forest is the natural vegetation over the whole of the inter-tropical area that has a dry season, with the exception of the very wet areas referred to above. South of the equator it occurs in eastern Java and the islands eastward to Timor. North of the equator it survives over great areas of Burma, Thailand, Indo-China and the Philippines. In the wet season this forest is not unlike tropical rain forest. The trees are not quite so tall, lianas are not so prominent and there is more undergrowth. The leaves fall and the undergrowth dies down in the dry season. Among the trees of commercial value are *teak*, *pynkado* and *padauk* in Burma and Thailand, and *sal* in India. Over large areas the proportion of teak to other trees is as high as 15 per cent, a circumstance that favours economic exploitation. This natural selection of one kind of tree is thought to be the result of occasional widespread fires which destroy all but the more fire-resistant saplings and seeds. Many of the extensive pure stands of teak seen in Burma today are fairly recent plantations on areas cleared of natural forest. On sterile laterized soils in South-east Asia, and especially in Burma, a poor variant of monsoon forest known as *indaing* is common.

Shifting cultivators are more numerous than in the tropical rain

(Photo—Government of India Tourist Office)

1. A crowded bazaar at Agra

Moghul architecture in India: the Taj Mahal at Agra

forest, with the result that more ground is cleared. Dense *bamboo* up to 40 ft. high is the principal growth on abandoned plots, and tall timber takes many decades to recover. In Burma and the hills of East Pakistan this destruction of valuable trees has been controlled. Cultivators are encouraged to plant seedlings of teak with their food crops. By the time the plot is abandoned the seedlings are well established and a rapid regeneration of forest is assured. In the Shan States of Burma, where clearing has been especially heavy and uncontrolled, downland of coarse grass is common.

Where soils have been severely eroded as a result of cultivation, both in the forests and in areas of permanent agriculture, often only *lantana* scrub will grow. This is a mass of prickly stems forming a tangle up to 15 ft. high. Though most common in areas of monsoon forest it is also found in tropical rain-forest country. Once established, it spreads rapidly and recovers quickly after burning. It is a serious menace in Burma.

There is no sharp division between the rain forests and the monsoon forests. A gradual transition occurs as one passes from areas with rain throughout the year to areas with a dry season. On the mainland this transition takes place between northern Malaya, where rain forest is characteristic despite the comparatively dry months of January and February, and Rangoon and Moulmein 900 miles farther north, where the dry season lasts for five months and the monsoon forests are fully developed.

Inter-tropical coastal forests. Hardly deserving the name forest in the popular sense of the word, *casuarina* trees are common on sandy coasts; seen from a distance they resemble cypresses. They form belts of open woodland less than 200 yards wide between the sea and the rain forest. In Madras they are much used for stabilizing sand-dunes on coasts exposed to the winter monsoon.

Mangrove forest quickly colonizes coastal mudbanks which are exposed at low tide. So mangroves are found wherever inshore waters are sheltered enough for mudbanks to form. They attain their most impressive development on sheltered coasts to which rivers bring ample supplies of alluvium. Such areas are the west coast of Malaya and the north coast of Sumatra, both well protected from the winter and summer monsoon winds. On the east coast of Malaya gales and high seas of the winter monsoon prevent the formation of mudbanks and there are no mangroves except in sheltered inlets.

Mangrove forests are tidal. On the seaward edge the tide range is slight, but it might build up to 15 ft. in the creeks on the landward edge. As sedimentation continues, the seaward fringe advances. On the landward side the older mangrove traps more mud and ground levels are raised above the tide. When that happens the mangrove is replaced in turn by freshwater swamp forest and ordinary dry-land forest. Thus mangrove forest is a stage in the natural reclamation of the sea. The mangrove belt in the strict sense is rarely more than 10 miles wide and usually much less. Taken together with the freshwater swamp, the overall width of badly drained country in southern Sumatra is nearly 100 miles. Mangrove forests contain plenty of stout timber in trees up to 100 ft. high and have the advantage of accessibility by water transport. They are extensively worked near ports for wharf timbers, firewood and tannin bark. The tortuous shallow channels are a hindrance to port development and navigation by large vessels.

Broad-leaved deciduous forests are native to the parts of the region which lie outside the tropics, principally China and Japan. There is a gradual transition northward from the monsoon forest, and the woods of southern China and Japan are semi-tropical.

Mountain forests. In the equatorial islands mosses thrive in the cooler and very damp atmosphere over 5,000 ft. They form thick blankets which hang from tree branches. Similar moss forest occurs in the eastern Himalayas above 8,000 ft. A spectacular feature in Yunnan and the eastern Himalayas is the rhododendron forest between 9,000 ft. and 11,000 ft. The rhododendrons grow as trees with straight trunks and large crowns, flowering magnificently in May and June. In Yunnan, the Himalayas, Japan and Manchuria there are considerable stands of coniferous timber.

Jungle. This term is applied in popular writing to all kinds of natural vegetation from rain forests to lantana scrub. It occurs frequently on the one-inch maps of Ceylon to describe the scenery. In India it is a common name for any rough and uncultivated country, and the adjective *jungly* is applied to the land and its inhabitants. In geographical literature the two words have no more precise a meaning than that.

5

Soils

A great deal about the processes of soil formation in tropical and equatorial areas still remains undiscovered; and an account of this subject without reference to the complexities of chemistry and biology is likely to be grossly oversimplified. However, the geographer with insufficient knowledge of physical science is on fairly safe ground if he works from the assumption that farmers everywhere will seek out the best soils for their purposes whenever there is a choice.

In Monsoon Asia it is the alluvial lowlands that have attracted the farmers. The plains of the Ganges, Hwang Ho, Yangtze and countless smaller rivers support densities of rural population well over 1,000 persons to the square mile. They are so crowded that if there were any better land available people would move to it. Yet neighbouring uplands such as the Deccan and the highlands of Shantung carry much lower densities. The fact of outstanding importance is that with rare exceptions the soils of alluvial lowlands give better yields than those of upland areas. A farmer can make a better living on an acre or so of lowland than on a much larger area of upland.

Soils of the inter-tropical areas. The superiority of the lowland soils is especially marked in the inter-tropical areas. The early European planters, impressed by the luxuriant vegetation on both lowland and upland, were puzzled by the rapid decline in yields of plantation crops in hillside clearings. The explanation lay in the quick deterioration of forest soils when the trees are removed. This is most easily understood if the upland soils are regarded as having two constituents: mineral grains from the decomposed rock and organic material, or humus, which forms a coating around each mineral grain. Most of the plant nutrients are in the humus: usually the mineral grains provide very little. Crops will flourish only so long as the humus content is adequate. Under forest conditions two processes govern the quantity of humus: first, the breaking down of humus by bacteria into soluble compounds which are leached from the soil by rain water; second, the replenishment of humus by the decay of leaves and dead plants on the forest floor. Where forest is well developed the second process is dominant and the humus content is maintained at a high level. When forest is cleared the supply of decaying vegetation is greatly reduced, while the destruction of humus is accelerated by the higher temperatures of the soil exposed to direct sunlight. Replenishment fails to keep pace with destruction and the humus content falls. The soluble plant foods are removed by soil water which seeps into streams, and are either reconcentrated together with river silt at lower levels in alluvial tracts or carried out to sea.

Another process which contributes to the poverty of upland soils is *laterization*. This is the process whereby, under conditions of high temperature and high rainfall, the complex silicates are broken down and removed by percolating ground water. The upper layers of the soil are left with a concentration of iron compounds and clay which imparts a red or purple colour. The more laterized a soil, the poorer it is for cultivation. Most inter-tropical soils are laterized to some extent. The end product of the process is *laterite*, a red earth useless for agriculture. It can be cut into blocks easily with a spade; the blocks harden on exposure and make a cheap and durable building material.

Laterization is a slow process. It is most advanced in soils which have been subject to it for long periods. Very old residual soils on Sunda Platform areas such as Malaya and Borneo, which have been continuously above sea level since Cretaceous times, are often highly laterized. In upland areas farther from the equator in India and South-east Asia, where a dry season coincides with cooler weather, the process is arrested for a period each year and the soils are less laterized. The most northerly soils showing signs of laterization are in hilly country immediately north of the Yangtze.

Laterization takes place in alluvial soils also, but those soils are generally so young that the process has not gone far. In many alluvial areas soils near rivers are continually renewed by fertile silt deposited by flood water. In older alluvium laterization is often apparent. For example, in East Pakistan patches of older alluvium projecting from beneath newer material are reddish in colour and less productive than the younger soils.

Taking the inter-tropical areas as a whole, the alluvial lowlands offer attractive farming country because their soils are young, and so free from laterization, and also because plant nutrients leached from the soils of high ground are to some extent reconcentrated in the soil water of lowland areas. This applies not only to the dry land of the alluvial plains but also to the coastal swamps. Drainage of those swamps is so expensive that it has been attempted only on a small scale, but the reclaimed soils are highly productive. Additional considerations are the flatness of the plains and the consequent ease of communication, water supply and irrigation.

Soils outside the tropics. In north and central China and Japan laterization does not occur. Other reasons must be sought to explain the value of the lowlands. The upland soils are liable to erosion, and they are often thin and inconvenient to work, on account of slope and exposure. The lowland soils are deep and easily worked. Some are frequently restored by silt during floods.

Uplands of exceptional fertility. In all uplands there are small areas of river alluvium and sometimes old lake beds and volcanic deposits which give good local soils, but there are two large areas of high ground in which the entire soil cover is good. These are in the highlands of Java and the loess uplands of north China. In both areas the humus content is low, especially so in the loess. In both instances the soils owe their productiveness and ability to support dense rural populations to their mineral particles. The Javanese soils are derived from basic volcanic ashes. These soils, rich in plant foods, are so recent as to be only slightly laterized—if at all. The loess of China is an accumulation of fine wind-borne material rich in lime.

The above paragraphs attempt to convey only a few significant aspects of a difficult subject. Their broad generalizations apply fairly well, but it is easy to find details of the landscape that contradict them. For example, terraced ricefields on poor hillsides in south China have been tended and manured with such care for generations that the soil is as productive as that of any lowland, while areas of rich alluvium along the Irrawaddy south of Mandalay have been reduced by erosion to badland topography clothed in lantana scrub.

6

Historical Geography

At the present time Monsoon Asia is divided for the most part into large states. Each is governed from a national capital, and has definite boundaries except in a few remote and mountainous parts where border territory is disputed. The scene is dominated by three political entities, China, India and Japan, which show every sign of retaining and enhancing their national status. Despite weak government in some of the smaller countries, the overall picture is of strong centralized administrations that are likely to endure. This has not always been so. For much of history the states shown on the political map of today did not exist. For example, there was in the political sense no India, no China, no Malaya. The region was divided into countless little states, some so small as to occupy only minor river valleys. Boundaries were unstable. Frequently a state would become strong enough to annex its neighbours and establish a bigger and more powerful unit that lasted until corrupt government and the avarice of the ruling house became intolerable and the population rose in revolt. The larger unit would then break up into several independent parts, or perhaps neighbouring princes would march in and divide the place among themselves. Occasionally, however, one ruling family would become powerful enough to extend its authority over hundreds of miles of country and establish for a time a centralized state comparable in area with the big national states of today. These large states, or empires, of the past are particularly interesting to geographers. The following paragraphs outline the circumstances of society and physical geography in which they came about.

For a state to maintain secure borders and centralized government, an army and a civil administrative service are essential. Soldiers and civil servants produce no food. They must be supported by other people who grow food surplus to their own needs. Therein lay the difficulty of the little states of Asia. As now, practically all their inhabitants were poor farmers who made a living by intensive agriculture and produced hardly enough to feed their own families. Taxes in the form of compulsory state service and a share of the annual food crop made life very hard. There were severe limits to the proportion of the total population that could be spared for occupations other than agriculture. The minute surpluses of many peasant families were necessary to support one man in public service. Clearly the states with the largest populations farming the best land had the greatest agricultural wealth and surplus manpower. Only the great alluvial lowlands could support populations large enough to provide the food and manpower to create and administer an empire. It is not surprising that the most successful empires were first established on the large alluvial lowlands and later extended their control over neighbouring poorer country. For some rulers there was the possibility of additional wealth from commerce. Some states of peninsular India grew rich enough on oversea trade in silks, spices and precious stones to employ European soldiers, but they were exceptional.

In the capital cities merchants and the upper classes lived well. There were many with leisure for art and learning. The literature and the mosques, temples and palaces that survive are evidence of high cultural achievement, but that was for the few and it was based on the unrelieved poverty of the agricultural masses. The kings and the numerous sons and others with claims to succession were often desperate people. Sometimes an eldest son followed his father to the throne in peace. More often, and especially among the Muslims, the death of a ruler was marked by treachery. Each claimant sought to eliminate his rivals. Blinding, flaying alive and trampling by elephants were accepted methods.

In the past the total population of the region was of course much less than it is now. The capital cities were very small by our present standards. In the Bronze Age intensive agriculture became well established. Then, and even much later, the small states were no more than islands in a sea of primitive hunters and herders at a neolithic stage of development. The peasant farmers gradually increased, and took over more land; the backward people were absorbed into the more advanced culture. The final stages of this assimilation are still going on. For example, some Deccan tribes and various tribes of south-west China are only now embracing the farming methods and social outlook that will ultimately make them Indians and Chinese respectively, in the cultural sense.

The Indus Civilization (2500–1500 B.C.)

About 3000 B.C., in the upland valleys of what is now Baluchistan and Afghanistan, there were peasants who grew wheat and other crops and knew simple methods of irrigation. Some of these people seem to have moved into the Indus lowland, an environment very different from the uplands. The rivers were wider and harder to control, but there were the advantages of better soils and a kinder climate. The newcomers to the Indus used these advantages by devising a social organization that ensured flood-control, drainage and irrigation on a scale vastly greater than any previously known. In addition to numerous towns, they maintained two great cities, Mohenjo-daro on the lower Indus and Harappa in the Punjab. Both cities had walled citadels. The lower part of Mohenjo-daro had a geometrical plan. Main streets were 30 ft. wide and even narrow lanes had brick-lined drains. Excavations at Harappa have revealed a granary with rows of brick platforms for pounding grain, and barrack quarters for workmen. Bronze implements and weapons have been found, as well as blades made of chert. There are also seal-stones, the largest about one and a half inches square, carved with human and animal figures, and a pictographic script no one can yet interpret. In all, the remains of about 60 towns and villages have been recognized. They form two distinct groups, one astride the Indo-Pakistan border along the abandoned channel of the Ghaggar (or Sarasvati) which was a great river flowing parallel with the Indus, and the other mostly within 50 miles of the lower course of the Indus. Mohenjo-daro and Harappa might have been capitals of two separate states, or there might have been one state over 1,000 miles across from the sea to the Himalayan foothills (Figure 6). In the highest levels in the ruins of Mohenjo-daro groups of human skeletons and other evidence of violence indicate that the city was sacked, presumably by some of the invaders who poured into India from central Asia at intervals whenever there was no power in northern India capable of stopping them. If the provisional dates suggested by archaeologists are right the Indus civilization flourished for about 1,000 years. Recent excavations at Lothal at the head of the Gulf of Cambay yielded evidence of foreign trade at this period: a dockyard 700 ft. long and a town of brick houses and shops.

The Mauryan Empire (322–185 B.C.)

When Alexander the Great invaded India in 326 B.C. he conquered several small states in the Punjab. At Taxila, the prosperous capital of the country

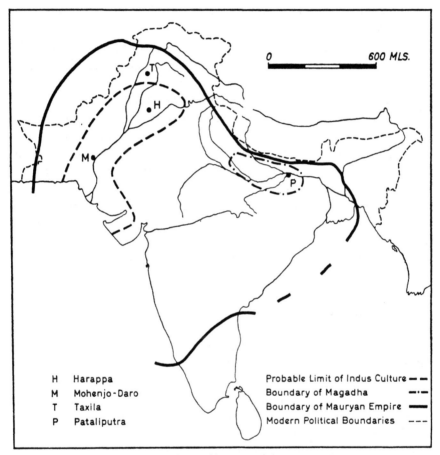

Figure 6. *The Indus civilization and the Mauryan empire*

between the Indus and the Jhelum, he was welcomed with gifts of 3,000 oxen and over 10,000 sheep. (Today the ruins of Taxila, twenty miles north-west of Rawalpindi, are an impressive sight.) After the refusal of his soldiers to advance beyond the Beas, Alexander withdrew. He died at Babylon in 323 B.C., and his empire was divided among his generals. One of these, Seleukos Nikator, marched into India in 305 B.C. to recover the Punjab. He was opposed and defeated by Chandragupta Maurya, king of Magadha, a state on the Ganges with a capital at Pataliputra (Patna). Chandragupta thus acquired the Punjab, the Indus lowlands and much of Afghanistan. This was the beginning of the Mauryan empire which was extended under Bindusara and Asoka to include peninsular India as far south as Mysore (Figure 6).

The powerful and efficient administration was controlled directly by the kings. Severe taxes were enforced throughout the empire. Where they were not collected directly by central government officials they were gathered by local princes who sent them as tribute to the paramount ruler. A well-equipped army of 700,000 was paid regularly, an unusual achievement in Asian states even in much later times. The walls of Pataliputra had 64 gates and 570 towers, and a moat filled by water from the Son. The site was a strong one in the angle of confluence of the Son and Ganges. Parts of the moat and timber palisade are to be seen today, but the river courses have changed and the confluence is now about 12 miles above the city. Buildings were mainly of timber, and though none survives there is evidence of their magnificence in the writings of contemporary Greek visitors. Sculpture reached a high standard. There were many towns and good roads. Rice was the principal crop, and irrigation was elaborately planned and maintained. Merchants traded in silk, muslins, ivory and gold with states of southern India and the eastern Mediterranean. In addition to the sea routes to Babylon and Suez, there was a land route from Taxila to Balkh and thence down the Oxus to the Aral Sea and to Europe. The whole cost of empire was borne ultimately by the peasants, for whom conditions were so oppressive that Chandragupta took constant precautions against attempts on his life. In his palace he never slept in the same room on two consecutive nights. Wherever he went he was attended by a bodyguard of Amazons. Yet it was a time of religious toleration: Hinduism, Buddhism and Jainism flourished.

The Gupta Empire (A.D. 320–500)

The Gupta dynasty also originated in Magadha. The territory of the first ruler, Chandragupta I, was limited to within 300 miles of Pataliputra (Figure 7). His successor established firm control throughout the Ganges lowland and sent armies into the peninsula. Although the conquered areas in the eastern Deccan and the Punjab were not directly administered, the local princes paid annual tribute to the Gupta king and acknowledged his paramount status. Later the country north of the Narbada as far as the coast of Kathiawar was occupied, and the acquisition of several small ports bringing a prosperous oversea trade with the west added wealth from customs duties.

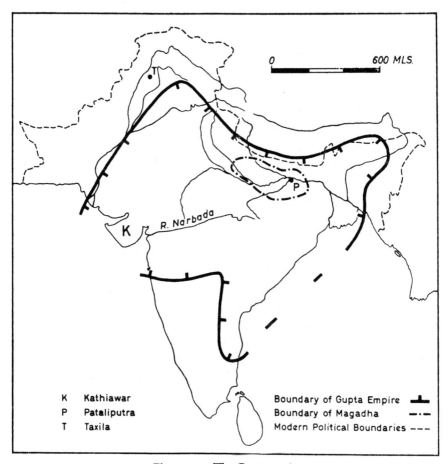

Figure 7. *The Gupta empire*

As in the Mauryan period, the empire held together only so long as the rulers were of exceptional personality and ability. From the writings of Chinese pilgrims who came by sea and by the oasis routes of the Tarim Basin and across the Pamirs to visit Buddhist monasteries in India, it is clear that the Gupta kings were able to govern effectively without the extreme discipline and oppression of the Mauryan regime. Music, architecture, painting and sculpture were of a high standard. Mathematics and astronomy were studied. The decimal system, which originated in India, probably dates from this time.

Towards the end of the sixth century the pressure of nomadic people from beyond the passes of the north-west overcame the resistance of Persia and other states in that area. Hordes of invaders came into India and the Gupta organization disintegrated.

The Sultanate of Delhi (A.D. 1206–1526)

Delhi first achieved outstanding importance when it became the capital of a succession of Muslim sultans. Several were adventurers who came at the head of armies from the country beyond the passes of the north-west. A minority among the people of India, the Muslims owed their rapid success to good martial discipline. They were remarkable for their extreme cruelty and religious fanaticism. Unlike earlier invaders, who were quickly assimilated into the Hindu and Buddhist society, they retained their separate identity as a ruling class. Their religion required them to convert others to their beliefs. Those who resisted were likely to be killed; often when Hindu rulers were defeated the civil population was massacred. A restraining influence was the dependence on the subject people for administration and public works, tasks for which the invaders had neither the ability nor the numbers. Most of the Muslims in India were not invaders but converts from the local Hindu population. Many were from the poorer peasantry, for whom Islam appeared to offer protection from the extortions of upper-class Hindus. By the fourteenth century the sultans' authority was recognized far south in the Deccan where it encountered a powerful association of Indian states, the empire of Vijayanagar.

Vijayanagar (A.D. 1336–1565)

Vijayanagar (Figure 8) claims attention as the largest stable political entity of peninsular India. Entirely Hindu, it prevailed for over 200 years. It included the whole of what is now Madras and Mysore. The capital, also called Vijayanagar, had a population of over 250,000. Government was oppressive. Yet—despite strong administration and cruel taxation to provide, among other things, a large standing army—the available wealth and manpower were slight compared with the resources of the Muslims whose base was in the richer plains of the north.

The Moghul Empire (A.D. 1526–1857)

To establish an empire in India a bold and enterprising outsider had no need to invade the whole country with troops, nor garrison every town and village. It was sufficient to strike quickly with a small force and gain control of such key positions as Delhi and Agra. The defenders were generally newly recruited and badly trained. However numerous, they were no match for tough, experienced fighters. Once a foothold was established in the lowlands, the wealth and manpower of the entire area became available for the ruling house to extend its power over the Deccan. This is illustrated by the careers of Babur and his successors, who created the most recent of the great Indian states based on autocratic rule, the Moghul empire (Figure 8).

On the death of his father in 1494 Babur (aged 12) inherited the small principality of Ferghana, a fertile lowland in the mountains at the headwaters of the Syr Darya which flows into the Aral Sea. A gifted leader, he took part in the continuous local wars. In 1504 he captured Kabul with fewer than 300 men. In 1523, with a force of 12,000, he invaded India. At Panipat, 60 miles north of Delhi, he defeated the last of the Delhi sultans—whose force numbered over 100,000 men and 100 war elephants —with artillery newly introduced from Europe. Other conquests followed. At his death in 1530 Babur, the first Moghul ruler of India, held an empire which included southern Afghanistan, the Punjab, the Ganges lowland almost to the delta and the northern fringes of the Deccan. Only the Punjab and the Ganges–Jumna *doab* were directly administered. The rest, apart from a few strategic positions, remained in the hands of local chiefs. Slowly, with much fighting and temporary reverses, a strong central administration was created to govern the whole area. By 1576 Akbar, the fourth emperor, held all the Indus and Ganges lowland except Sind, the country southward to the Narbada and northward to the Hindu Kush. He died in 1605 without achieving his ambition to control the Deccan and recover the ground held by his ancestors in central Asia.

When Aurangzeb, son of the sixth ruler, Shahjahan, became viceroy of the Deccan, in 1636, only the country south of Madras and Calicut was outside the Moghul domain. In 1690, when Aurangzeb was emperor, even Tanjore and Trichinopoly were paying tribute to Delhi. Most of the

Deccan however was relatively poor country, and as a whole it was a financial liability. After the death of Aurangzeb, in 1707, government deteriorated because none of the kings was capable of infusing new life

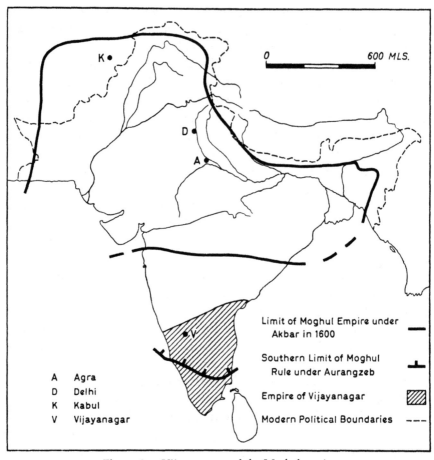

Figure 8. *Vijayanagar, and the Moghul empire*

into the administration. The Deccan states broke away. The last Moghul emperor was deposed by the British at the time of the Mutiny, but there was no empire worthy of the name after 1707.

Much of the architectural wealth of Agra, the early capital, and of Delhi, to which the capital was transferred in 1648, is of the Moghul period.

The Han Dynasty (206 B.C.–A.D. 221)

In China there were five periods of strong central government when civilization flourished, separated by times of political upheaval and insecurity. Primitive cultivators and hunters were widespread in the neolithic period. By Bronze Age times intensive agriculture with irrigation was practised in the fertile loess uplands of the north, and the people of the Mongolian grasslands had devised their specialized nomadic economy with domesticated herds of sheep, goats and horses. As in the Indus area, the permanent extension of agriculture to the great plain awaited the evolution of social organization. Only in disciplined states could sufficient labour be relied on to control the river floods. Gradually several independent agricultural states came into being in north China. They suffered from the raids of their nomadic neighbours and protected themselves with earthen ramparts along their northern borders. Towards the end of the third century B.C. Shi Huang Ti, the chief of one of those states, became strong enough to impose his will on the others and form the earliest united empire of China. As first ruler of the Han dynasty, he created a strong administration. He raised an army large enough to improve and extend the separate lengths of rampart, so forming the Great Wall of China which runs for 1,600 miles from the eastern edge of the Tibetan plateau to the Gulf of Chihli (Figure 9). The cultivators had already settled northward to limits beyond which the rainfall was usually insufficient for intensive agriculture, so the Great Wall marked the frontier between cultivators and nomads. Today the wall is an impressive structure of stone with towers at frequent intervals, but this masonry is a late improvement. The Han wall was of earth with timber palisades. Its construction and maintenance were enormous tasks for a state dependent on the resources of a peasant population.

From their early base on the northern plain, later Han kings controlled the country southward beyond the Sikiang. The whole of that area became one state, with most of the people converted to intensive agriculture, but the north retained its position as the most densely populated part of the empire. The Han kings were also interested in trade with the west. They garrisoned the oases along the trade routes through the Tarim Basin. At one time they claimed authority as far as the Caspian

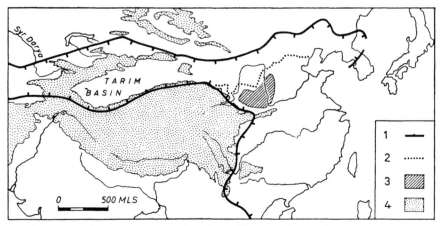

Figure 9. *The Han empire* Explanation of key: 1—limit of Han empire at its maximum extent; 2—the Great Wall; 3—the loess highlands; 4—land over 9,000 ft.

Sea, on the farther shores of which lay the outposts of the Roman empire.

The Tang Dynasty (A.D. 618–907)

The Tang empire was smaller than the Han: unruly tribes in the southwest were not thought worth holding down, and the land routes through central Asia had declined in importance with the increasing use of sea routes to India and the west. The oases of the Tarim Basin were left to Muslims who invaded from the west. By contrast, the southern provinces were more densely settled, and the north had become so heavily populated that rice was imported from the south to supplement local supplies of wheat and millet. The absence of a natural north-south waterway for this had long been a hindrance, but the immediate predecessors of the Tang rulers improved and lengthened existing canals to form the Grand Canal from the Yangtze to the Hwang Ho (Figure 10), and this was the economic lifeline of the Tang state. The capital was at Sian.

The Sung Dynasty (A.D. 960–1127)

The approximate extent of the Sung empire is shown in Figure 10. The area of the lower Yangtze surpassed the north in density of population and in total numbers. This remarkable development of the south followed

3. Moghul architecture in India: the Red Fort at Delhi

(*Photo—Government of India Tourist Office*)

4. Modern architecture in India: Marine Drive, Bombay

the revolutionary changes in Chinese agriculture that took place about A.D. 1000, and included double cropping. Two crops a year can be grown on the same land only if each occupies the ground for a short enough period, but the ordinary strains of rice require up to 200 days in the soil, and the remaining 165 days fall at the least favourable time of year for other kinds of agriculture. This difficulty was overcome by introducing strains of early-ripening, drought-resisting rice, known as *Champa* rice (after the state in what is now southern Vietnam), which needed only 100

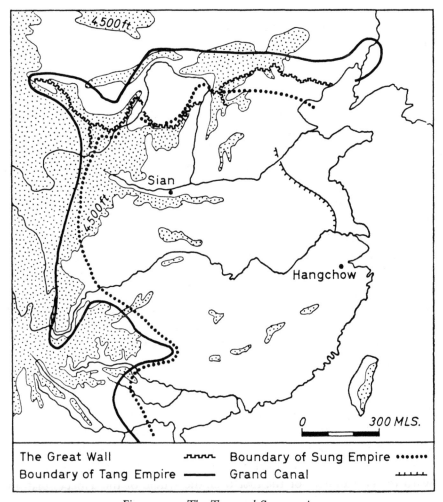

Figure 10.   *The Tang and Sung empires*

days in the soil. In theory three crops of rice a year became practicable wherever climate and irrigation were suitable, but in fact the new rice did not yield quite as well as the old. It was more profitable to have either one rice crop followed by wheat, or two rice crops followed by vegetables. Champa rice became the basis of many local varieties, enabling the Chinese to spread their agriculture from the best soils of the lowlands into the harsher conditions of the surrounding hills. The drought-resisting qualities meant that they could also grow rice farther north, though the winter climate of north China is too severe for double cropping.

For the first time in Chinese history the capital was in the south, at Hangchow.

The Mongol Dynasty (A.D. 1280–1368)

In the eleventh century the Mongol tribes of the Mongolian grasslands rose to military supremacy under Genghis Khan and his grandson, Kublai Khan. They amassed an enormous empire that reached westward to the Mediterranean and southward to Annam and Burma. For his capital, Kublai Khan chose Peking. The Mongol part of that city is unlike other Chinese towns in that it is laid out on the plan of a nomad camp. There are wide avenues on a rectangular pattern, the whole surrounded by a wall 50 ft. high.

The Mongols were good at fighting but neglected administration. Communications were poor and the empire soon began to fall apart. Kublai Khan's successor was a Buddhist. He ruled only Mongolia, China and the conquered lands of South-east Asia. By that time the western Mongols had been converted to Islam and would not recognize an infidel king. In China there was great oppression; ultimately a new line of Chinese leaders, the founders of the Ming dynasty, drove the Mongols out, and in 1368 Peking was restored to Chinese rule.

The Manchu Dynasty (A.D. 1644–1911)

The Ming emperors consolidated their authority throughout China, but recovery of former Han possessions in central Asia awaited the line of kings who originated in Manchuria. Under the Manchus, Chinese suzerainty was extended northward to the border of Siberia and westward to the Pamirs. For the first time the whole of the Tibetan plateau was claimed, and Chinese troops and political agents were stationed at

Lhasa. But in the later stages the Manchus sadly failed to adjust themselves to modern political thinking. Rebellions were frequent. In 1911 they were overthrown by the Nationalist régime.

SOUTH-EAST ASIA

Throughout South-east Asia many small states existed from an early date. By the fourth century they were in cultural contact with India and Buddhism had spread over the whole area. Of the larger units, the seventh-century state of Srivijaya in southern Sumatra controlled the Strait of Malacca and part of the mainland. It was conquered in the eighth century by the Sailendra kings of Java, who dominated most of Indonesia for 300 years. In the fourteenth century the Majapahit empire of Java was the principal power. Ruins of religious monuments in Java and of the walled city and temples of Angkor, capital of the thirteenth-century kingdom of Cambodia, are evidence of advanced culture, but none of these states attained the size and political organization of the big empires of India and China. Probably the most important single factor operating against the rise of large political units in South-east Asia was the absence of a large and well-placed lowland. Such lowlands as exist, and they all offer good agricultural opportunities, are comparatively small and scattered. None was extensive enough to provide surplus manpower, so that no ruler could dominate much more than his immediate surroundings.

POLITICAL ASSOCIATIONS WITH EUROPEAN POWERS

Trade in spices, silk, muslins, ebony, sandalwood and jewels from Asia to Europe dates back at least to Roman times. The land and sea routes from Asia converged on the eastern Mediterranean. Cairo, Alexandria, the cities of the Levant and the towns at oases on the desert routes all grew rich. In the Middle Ages several Italian republics acquired great wealth from this commerce; supreme among these was Venice, which for a time had a virtual monopoly of trade in eastern goods moving farther into Europe.

The first to seek alternative ways to Asia and so break the Venetian monopoly were the Portuguese. With their fast, square-rigged, flat-bottomed caravels, the best ships at sea in those days, they pioneered the route round Africa to the Indian Ocean. In May 1498 Vasco da Gama anchored near Calicut. Opposition of Arab traders on the Malabar coast was overcome. Permanent trading stations were set up there and in Ceylon, Malaya and at several places in Indonesia and on the China coast at Macao. The profits were high, not only from goods carried to Europe but also from the trade between the Asian countries themselves which was taken over from Arab and Chinese shipping. The Portuguese had things much their own way for a hundred years, until the Dutch appeared as strong competitors, shortly to be followed by the British with the founding of the East India Company in 1599. France and Denmark were also early in the field, and Sweden, Austria and Prussia took a hand in the eighteenth century: of these only the French efforts had much success.

Gradually much of Monsoon Asia passed under European control. Seen in human terms, these power politics make fascinating reading. From the early Portuguese captains and William Hawkins, who sailed to India in 1607 under instructions from the British East India Company that there was to be no swearing or dicing on his ships, to the eighteenth-century battles of Clive and Dupleix, the Indian Mutiny and other nineteenth-century events, the story abounds with personalities.

By the beginning of the present century, British rule was established throughout India (except for the small coastal possessions of the French and Portuguese), Ceylon, Burma, Malaya and part of Borneo. The Dutch held most of Indonesia, while the main French possessions were in Indo-China. Of the large countries, only Siam (Thailand), Japan and China were independent. In China several European powers had special trading rights and legal privileges in coastal cities. The Philippines, after three hundred years of Spanish rule, became a dependency of U.S.A. in 1898.

7

Agriculture

THE MAIN FOOD CROPS

Rice is the principal food grain of Monsoon Asia. Throughout the hotter, wetter areas of the south, including southern Japan, the Yangtze and Ganges lowlands, much of peninsular India and almost all South-east Asia, the alluvial land presents a monotonous pattern of rectangular paddy fields. Where the rainfall is adequate for rice, but restricted to a sharply defined wet season, the main rice crop is planted so that it can grow when the fields are flooded by rain water (often supplemented by natural flood water and more rarely by irrigation), and the harvest takes place between November and January after the water has drained away. In Malaya and much of Indonesia, where dry seasons are short or non-existent, some rice is always ready for reaping, though most of the harvesting is arranged to coincide with the relatively dry months. In Malaya those months are between October and March when the days are shortest. The annual range in the duration of daylight is slight, only nine minutes at Singapore, increasing to 37 minutes at Alor Star, but many kinds of rice are sensitive to that range and will flower only in the period of shorter days. Good crops are raised on terraced hillsides where individual fields might be as narrow as four feet. Even in the drier parts of these southern areas it is a popular crop wherever irrigation water is available.

Several thousand varieties of rice are grown. There are high-yielding strains with growing periods of about 200 days, others not so productive but maturing in 120 days or less and yet others which are drought resistant. In Japan and at the famous research station at Cuttack on the Mahanadi delta, new varieties resistant to drought and disease are constantly under investigation. Generally the crop grows in six to eight inches of water and stands three to four feet high at maturity. Locally in Malaya and Thailand there are varieties that keep pace with rising flood water by growing over two feet a day: small areas of rice mature in as much as 15 ft. of water and are harvested from boats. Hill rice is the name for varieties grown in dry fields on hillsides: these are relatively un-important.

North of the Yangtze lowland, southern Korea and southern Japan, cooler and drier conditions favour other crops. Some rice is grown in Manchuria and the oases of the Tarim Basin. Even in Hokkaido, the northern pioneer fringe of Japanese settlement, rice is more important than any other grain crop.

Wheat is a winter crop in India and Pakistan. It is grown in the western Deccan as far south as the Kistna river, but it is most prominent on irrigated areas of the Punjab and in the western half of the Ganges lowland. In the Yangtze valley wheat occupies some of the drained paddy fields in winter. Farther north it is entirely a summer crop; autumn-sown over most of the plain of north China and in southern Japan; but spring-sown in the harsher climate of the hilly country near the Great Wall and in Manchuria.

Millets, of which over 30 species and several hundred varieties are grown, are characteristic of the more arid areas. In India *jowar* (great millet) and *bajra* (bulrush millet) are widely cultivated, and other types are important locally. Among the several millets grown in China, *kaoling* is especially prominent in the north. From the nutritional viewpoint millet is superior to both rice and wheat as a human food, but in most areas it is associated with poverty and despised by the richer people. Rice is the food grain in greatest demand, despite the deficiency diseases such as *beri beri* associated with a rice diet unrelieved by vegetables, other grains and animal products. The stalks of millet, which often grows to 10 ft., make good animal fodder and light building material.

Other food crops are edible oilseeds such as groundnuts, sesame and mustard. There is a great range of pulses: among them gram, lentils and various beans, and small acreages of barley and maize. Vegetables and fruits, though dietetically important, occupy only very small acreages.

Crops grown for sale in local markets might include some of the food crops described above. Usually a peasant family grows only enough for its own needs, but it might occasionally sell some of its grain or vegetables to buy meat for rare and special feasts. In a few areas there are substantial annual surpluses for disposal. For example, rice from Burma and Thailand is in demand to feed plantation labourers in Malaya and Ceylon. Jute in the Ganges delta and Assam and sugar cane in northern India are important cash crops. Others are cotton, industrial oilseeds, tobacco, spices and sometimes small plots of rubber and coconut.

LIVESTOCK

Cattle, sheep, goats, pigs and chickens abound in Monsoon Asia, but except in the rare instances of dairy farms near such cities as Bombay, Delhi and Calcutta there is no animal husbandry as understood in western countries. Cattle are used everywhere as draught animals. They pull carts on the roads and ploughs in the fields. In India they are regarded by Hindus as sacred. A Hindu farmer will not kill a cow, though in some areas he might sell aged animals to Muslim neighbours for slaughter. In densely settled areas so high a proportion of the land is needed for food crops that very little can be left for pasture. Nor is there normally any room for fodder crops. The cattle, which include the great black water buffalo of the wetter areas, are generally in poor condition. Breeding is not controlled. Most of the fodder is given to the working bullocks, so that cows often have to live as best they can on stubble and roadside grass. In these circumstances cattle contribute little to the food supply. Where fodder crops are grown, as in some of the better-planned villages of the Punjab, dairy produce is important. Water buffalo are the main constituent of the carefully managed high-yielding dairy herds which supply the limited market of prosperous people in the major cities. The meat of sheep and goats in small amounts can be bought in local markets throughout India and Pakistan, and pork in Chinese villages. The ubiquitous chickens are small and tasteless. In the Mongolian grasslands animals contribute directly to the food supply, but the number of people

71

dependent on them is an insignificant proportion of the regional population.

In the equatorial zone rainfall is more than adequate for most of the crops. On the low-lying rice lands drainage is usually the problem, not water supply, though short dry spells often result locally in poor rice yields. Complicated irrigation systems bring water to terraced ricefields on the hillsides, especially in Java. Irrigation is most developed in areas farther north which experience a dry season and where the monsoon rainfall is low or unreliable. Enormous areas are watered by modern perennial canals in West Pakistan, northern India and parts of peninsular India. There are small canal systems in the dry zone of Burma. Elsewhere, and especially in peninsular India, earthen walls have been built across minor valleys to impound water. These reservoirs are called tanks. The smallest might extend for about 100 yards with a retaining wall no more than five feet high. Larger ones are several miles long, impounded by embankments rising to 50 ft. Further forms of irrigation include wells from which water is hauled in leather buckets by oxen (the principal method used in China), and modern tube wells with electric pumps. In the Tarim Basin and the north-west borderlands of Pakistan small areas are watered by *karez*, tunnels which run from catchment areas in screes high up on mountainsides and emerge at the edges of fields on the valley bottoms. The tunnels, some several miles long, are marked at intervals on the surface by vertical shafts surrounded by spoil heaps.

CROP YIELDS AND NUTRITION

The United Nations secretariat in Bangkok has collected figures of crop production throughout Monsoon Asia. Those figures have been converted into indices which allow for the different calorific values of standard weights of the several crops. A summary of the United Nations findings is given in Table 2. Food production (in terms of calories) per head of population for Monsoon Asia as a whole has recently been holding steady at about 14 per cent below the average level for the years 1934 to 1938, whereas the corresponding figures for the entire world are running at 3 per cent to 6 per cent above the pre-war standard. This is a serious condition in a region where nutritional standards were miserably low at the outset. The principal reason for the reduction of 14 per cent is the high birth rate. The population rises steadily year by year, but food production has lagged behind.

Taking the several countries of Table 2 in turn, however, the lowest indices are not necessarily associated with the worst conditions. The

TABLE 2

INDICES OF FOOD PRODUCTION PER HEAD OF POPULATION

(Based on the calorific value of the several kinds of food produced)

The average for the years 1934 to 1938 = 100

	1954-5	1955-6	1956-7	1957-8
The World	103	105	106	104
Monsoon Asia	86	86	88	86
Burma	71	71	75	65
Malaya	76	75	80	78
India	95	96	96	93
Pakistan	96	88	95	92
Ceylon	101	104	97	98
Japan	92	108	104	109

(No separate figures available for China)

Source: United Nations, Bangkok

Burmese, with an index only around 70, are comparatively well off. Before the war they had a very large rice surplus for export. The lower post-war figures mean they have less for export—not that the Burmese get less to eat. Throughout the last 60 years Malaya has imported rice to feed the labourers on the plantations. The lower post-war indices there indicate a larger food deficit now than 20 years ago, but Malaya can easily finance imports for the comparatively small population from the profits of tin and rubber. Things are not so easy in India and Pakistan, higher indices notwithstanding. Those countries were worse off at the start; they have enormous populations to feed and very little in the way of exports to pay for food, even if food in sufficient quantities were available. Ceylon is another importer of rice. There again the population is small and the imports can easily be paid for out of the profits of plantation crops: the comparatively high indices are the result of government encouragement of peasant food production. The high figures for Japan reflect the efforts of the Japanese to reduce their dependence on imports.

If the food production per head of population for the *whole region* suffers a further major decline the inevitable result will be widespread famine and millions of deaths from starvation. Almost certainly the published figures of actual crop production are inaccurate. Some go so far as to say that if the figures were true most of the people of Monsoon Asia would already be dead. Farmers throughout the world are not given to over-estimating their crops when they report to local officials, for fear of attracting higher taxation. But even if the yearly estimates are too low the *trends* indicated by the annual indices in Table 2 remain valid, and there is plenty of other evidence of hunger. This problem of food supply was recognized over 50 years ago by the western countries with responsibilities in the region. Much was done to improve irrigation, agricultural methods and food distribution. Since 1945 remedial measures by the Asian countries themselves have been backed by the resources of United Nations. Progress is being made along the following lines:

1—extension of the total cultivated area;
2—extension of irrigation;
3—provision of more natural and artificial fertilizers and better seed;
4—reorganization of farm holdings.

Extension of the total cultivated area. The Indian government has sponsored new farming areas in Assam. The Ceylon government has established new agricultural colonies in neglected areas. But both these schemes are minute in relation to the regional problem of food supply. Practically all the best land is already under cultivation. Extension of the total cultivated area cannot be expected to achieve more than purely local gains, at any rate in the foreseeable future. A useful gain would be the reclamation of the great coastal swamps of Sumatra, Borneo and New Guinea which are potentially highly productive. Such reclamation would be very expensive but might turn out to be essential.

Extension of irrigation offers greater rewards because higher yields (including two or more crops a year where only one is grown at present) can be expected from good land already in cultivation. Big perennial canal schemes are under construction in the Indus and Ganges lowlands, and in north China. Most of the present irrigated areas of the region are watered by wells and tanks which often fail at critical times: the replacement of the least efficient of these older works by modern canals fed from large reservoirs and tube wells would certainly remedy matters.

Provision of more natural and artificial fertilizers and better seed would bring immediate improvements: On experimental farms in India, where

TABLE 3

CROP YIELDS: 1962

(tons per hectare)

	Rice	*Wheat*
Burma	1·7	—
Cambodia	1·0	—
Ceylon	1·9	—
China	3·0	0·9
Taiwan	3·2	1·7
Malaya	2·3	—
India	1·3	0·7
Indonesia	1·7	—
Japan	4·9	2·7
Pakistan	1·4	0·8
Philippines	1·2	—
Thailand	1·3	—

Source: United Nations, Bangkok

oil cake and sulphate of ammonia are used, the rice yields are often five to six times those of the ordinary cultivator who uses very little manure. In India it is traditional to burn animal manure as a domestic fuel after it has been spread on the outer walls of buildings to dry: over half the farm-yard manure is probably lost to agriculture in that way, and artificial fertilizers are generally scarce and expensive. In contrast, the farmers of Japan and China prepare compost from animal manure and human excreta. The Japanese also use substantial amounts of chemical fertilizers, and the benefit to the fields is reflected in the higher yields of those two countries, recorded in Table 3; the rice and wheat yields of Japan are respectively four and three times those of India. But the present output of fertilizer from the new chemical works of India and China, though

valuable, is not enough. Rural education and higher standards of distribution are necessary if farmers are to realize the value of better strains of
seed.

Reorganization of farm holdings. Individual farms are generally small,
too small to allow the farmer a substantial margin of profit. In the rice
areas of south China the average holding is two acres. In the drier, less
productive parts of India many holdings are about 25 acres, but the
average for the whole country is under five acres. Only rarely is a farm
one piece of land. Usually it consists of five to ten separate plots scattered
throughout the village lands, as the result of inheritance laws which have
permitted each heir to claim a share of each of his father's fields; each field
is divided every time the ownership passes from father to sons. In one
village in Bihar an area of cropland 14½ acres in extent contained 59 plots
in 1912. By 1955 there were 92 plots. There are no hedges or fences. The
boundaries are uncultivated baulks about a foot wide: when fragmentation reaches an advanced stage the baulks take up a large proportion of
the land available to agriculture. If the traditional inheritance laws were
the only influence on the size and fragmentation of holdings most of the
ground would by now be taken up by boundaries and there would be no
room for crops. But when plots become very small they are sometimes
sold or rented to the holder of the next piece, and two or more plots are
then worked as one.

In some areas the reorganization of whole villages is encouraged by
state governments. The land is surveyed accurately and distributed among
the villagers in proportion to their original holdings. Each new holding
is one piece of land, often contour-terraced to reduce erosion. This reform
is hard to achieve because it requires the co-operation of all the land-
holders, but it is worth while; by doing away with many old boundaries
the crop area is increased, and cultivation becomes more efficient because
the farmer's attention is no longer divided between several scattered plots.
It also does away with the endless litigation over rights to small fragments.
Many villages in northern India have been improved in this way, and
there are new laws to prevent the subdivision of plots. In China collective
farming has been adopted to remove the worst effects of the old system.

It remains to be seen whether the methods described above will solve
the food problem. The present balance of opinion is that they will be

effective only if the populations concerned practise in addition a policy of family limitation. Once food production is increased to satisfactory levels the main hindrance to further improvement in living standards will be the size of the rural population dependent on agricultural employment. All farm work except ploughing is done by hand. The total labour available is needed only at times of great activity such as planting and harvesting. For the rest of the time many people are under-employed or unemployed, producing little or nothing, but consuming food. In other countries mechanization enables small agricultural populations to run the land with great efficiency. That solution is not yet practicable in most of Monsoon Asia for the following reasons. Machinery is expensive. Individual farms are too small to justify tractors and a full range of implements for each (a partial solution here is the provision of machinery which farmers can hire from village depots). Fields are generally too small for tractors, and no machinery suitable for wet-paddy cultivation is yet in mass production. Widespread mechanization would result in the unemployment of scores of thousands who now find a poor living on the land. There would be no future for them unless an industrial revolution were to provide factory employment in the towns. This is appreciated by the Asian governments. Everything cannot be done at once when resources are small. Large-scale industrialization must come, but it cannot flourish until the rural areas are rich and productive enough to provide a market for the factory products and produce surplus food for the new urban populations. The futures of agriculture and industry are thus closely linked.

SHIFTING AGRICULTURE

Shifting agriculture is the only form of cultivation in several of the remote uplands of South-east Asia. In Indonesia and the Philippines cassava, sweet potatoes and yams are grown, while in Burma and Assam maize and other cereals are also raised. The system is wasteful of good forest and supports only low densities of population. For the most part, central governments do not interfere. In Assam, where destruction of forest has resulted in floods in the Brahmaputra drainage, the Indian government met with armed resistance from hill tribes when it attempted to resettle them in permanent villages in the valley bottoms.

Plantation agriculture began to flourish towards the end of the nineteenth century to supply the new demands of European markets. The main crops are rubber, tea, oil palm, tobacco and coconut. Coffee and tung oil are also grown. The climate is suitable for these crops over large areas of Monsoon Asia, but plantations require substantial capital investment by individuals from western countries, and they came into being in large numbers only in those countries (directly controlled by European states) where business men considered their money safe. Although plantations are found in many parts of the region, there are five principal areas in which the landscape has been transformed by this method of agriculture: western Malaya, parts of Sumatra and Java, Ceylon, Assam and north Bengal, and the hills of southern India. Apart from the law and order associated with colonial rule, these areas offered no special advantages. Communications were poor in the forested areas where the plantations were established. Often port facilities were bad. Because the peasant populations were not greatly attracted by the low wages offered little local labour was forthcoming. Roads, railways and ports were all constructed with western capital. Labour was recruited mostly from southern India. Some Chinese also undertook work on plantations.

Even after the more urgent of the problems of large-scale cultivation under tropical conditions had been overcome, plantation agriculture remained a notorious financial gamble. Production took place in areas far from the ultimate markets, so that it was impossible for growers in Malaya and Ceylon accurately to predict the demands of European and American markets. Up to 1939 the annual output of plantations was generally greater than the demand, prices were low and profits at best were moderate.

Rubber is easy to grow and became popular with Malayan peasants who set aside parts of their land for rubber trees. They contributed a substantial proportion of the Malayan rubber output up to 1939 and still do so. They tap their trees only in years when the price of latex is attractive. When prices are low they neglect their rubber, without hardship to themselves, and give all their time to rice. On the other hand, if the estate workers are not paid and supplied with food, they disperse, and cannot easily be assembled again when prices improve. So the overheads of plantation agriculture continue whether prices are good or bad. Some have gone so far as to suggest that the peasant farmer could become the more efficient in producing rubber, and that the plantations might fail to compete. However, since World War II, and despite the competition of synthetic rubber, the market has been large enough to take both the plantation and peasant contributions. The indications are that the world will buy all the natural rubber it can get, provided that growers maintain high efficiency and continue to replace old trees with new high-yielding strains to keep prices down.

Tobacco is also grown by small farmers, but high-quality leaf comes only from the best Sumatran plantations. Tea and oil palm need more skill in cultivation than peasant farmers are generally prepared to give. They are exclusively plantation crops, still largely under European management, though Asian managers are increasing in numbers. Coconut plantations, on the other hand, have always been a field of indigenous capital and management. Two great advantages of the plantation system are the ease of ensuring uniformly high quality of the crop and the provision of funds for research into improved methods of cultivation. The results of the research are of course equally available to the small farmer.

8

Manufacturing Industry

Iron, steel and heavy engineering industries exist only in Japan, China and India. In 1957 the crude-steel production of those countries was 12·6, 5·3 and 1·7 million tons respectively. Together they produced hardly as much as the United Kingdom, and only one sixth as much as U.S.A. In all three the strenuous efforts of recent years to build new steel works and improve old ones are beginning to show results in higher output. The heavy engineering works are at the iron and steel centres and some of the major ports. Several countries produce small quantities of chemicals, but the larger units of the industry are limited to Japan, China and India.

In Monsoon Asia the spinning and weaving of cotton on power-driven machinery in factories has been the relatively easy first step in industrial progress. Such machinery is cheap, the skills easily learnt and the local market is usually adequate. The cotton industry has been established in small units in most of the countries. In Ceylon, one of the few still without a textile industry at the end of World War II, cotton mills are at last being set up with Japanese help. Large concentrations of the cotton industry in big factory units are limited to Japan, China, India and Hong Kong, and of these Japan and India are the world's largest exporters of cotton goods. Other textile industries are the processing of jute in Bengal, and the small woollen industries in north India, West Pakistan and China. Several countries produce silk. Japan also produces textiles from various artificial fibres.

Oil refining, limited until recent years to the oil-producing countries of Japan, Burma, Borneo, Java and Sumatra, is spreading to ports such as Bombay, Karachi and Manila, which command big local markets. China has several refineries.

There is a vast range of other industries, including cement, paper, printing, leather, food-processing, non-ferrous metals and light engineering. In Japan these are part of an industrial structure as comprehensive and complex as any in the west. Elsewhere they are usually in small units, many employing only 50 to 100 workers. Except for cement, which is made where the limestone is quarried, they are carried on mostly in the larger towns and ports.

Factories are not entirely lacking in the countryside. There are thousands of small rice mills, many at riverside villages to which rice is brought by sailing boats. In plantation areas it is generally convenient to do at least some of the preparation of the raw product before despatch to regional markets such as Singapore, Calcutta and Colombo. All the processes in the production of commercial tea are completed in air-conditioned factories on the plantations. Buyers in western markets need only blend the cured leaf from several sources. Rubber and oil-palm plantations have factories for initial stages of preparation. All these works are widely dispersed, generally away from main roads, and are therefore easily overlooked. In aggregate they employ a substantial number of workers.

Domestic and small-scale industry. Spinning, the weaving of textiles on hand looms and a great variety of other crafts occupy rural households at slack times in the agricultural round. No one knows how much is produced by these family activities. In a few countries they have been assessed together with small workshops under the heading of cottage and small-scale industry employing less than 20 persons. The proportion of the total industrial product contributed in 1955 by establishments in that category was as high as 76 per cent in Pakistan, 52 per cent in India, 32 per cent in China, but only 14 per cent in Japan, technologically the most advanced country in the region. Some of the handicrafts such as metal-working involve great skill and secure high profits in international markets. The majority serve only local markets. Spinning and weaving by hand is everywhere at a disadvantage in competition with local factory products. Consequently, it is declining throughout the region— except in India, where it has been officially encouraged as an aid to rural self-sufficiency.

Stages of industrial progress. A fair indication of a country's industrial development is the amount of power consumed per head of population; this quantity is not easily calculated, but the per capita production of coal and electricity for the selected countries in Table 4 gives a useful approximation. For comparison, data for U.S.A. and four European countries have been added. The figures for U.S.A. are outstanding. Even Denmark, the least industrialized of the four European countries, has a higher per capita production of electricity than any Monsoon Asian country except Japan. The electricity figures for Japan approach western standards. Next, but very far behind, come Singapore, Hong Kong and Taiwan. In the first two of these, enterprising Chinese and others have taken advantage

TABLE 4

COAL AND ELECTRICITY PRODUCTION PER HEAD OF POPULATION: 1960

	Coal (tons)	Electricity (kilowatt-hours)
Cambodia	—	10
Ceylon	—	27
China	0·5	6·2
Hong Kong	—	332
India	0·1	36
Japan	0·5	1,020
Malaya	0·01	171
Pakistan	0·01	11
Singapore	—	380
Taiwan	0·3	310
U.S.A.	2·5	4,100
Belgium	3·3	1,380
Denmark	—	780
France	1·3	1,360
United Kingdom	4·5	1,830

of stable government and commercial freedom to establish a multitude of industries. At Hong Kong this has been achieved despite the complete lack of local raw materials. In Taiwan the early start given by the Japanese has been maintained by the Nationalist Chinese with American help.

The size of the industrialization problem can be judged from the very low figures for India, Ceylon, Pakistan and Cambodia. To achieve industrial development comparable with that of Belgium the present Indian coal production would need to be 33 times larger and the electricity output over 38 times larger. Throughout Monsoon Asia it is recognized that agrarian poverty and dependence on capricious world markets in a few plantation products can be overcome only by industrialization. Each country has comprehensive development plans embracing agriculture and industry. Except in Japan, where progress is a relatively simple matter of building upon the great industrial achievements of the last hundred years, the difficulties are enormous. Capital and technical knowledge at managerial level can be partly provided from outside, but the skilled labour can come only from the rural masses. More elementary education in reading and writing as a foundation for technical training is vitally needed. Better banking, credit and marketing facilities will also help.

Japan is well placed to supply capital goods such as machine tools, but as industrial progress becomes general there will be an increasing risk of economic chaos resulting from conflicting national interests. Already the Japanese and Chinese are in keen competition in the markets of Southeast Asia. Up to a point that is a good thing, but unless regulated by regional trade agreements it could get out of hand to the detriment of all.

The multi-purpose hydro-electric schemes. Of all the countries of Monsoon Asia, only China has rich reserves of coal. Vietnam has plenty of anthracite. The Japanese resources are poor, and in India the coal is being used at a rate which is expected to produce a critical situation within 60 years. The coal resources of the remaining countries are negligible. Oil is widespread in association with the Alpine fold zone. Probably some oil-fields have yet to be discovered, but the known reserves are being increasingly drawn upon and supplies for the future are uncertain.

For these reasons fuel is regarded as a major problem, and in several countries the opportunities for hydro-electric schemes have already been studied. In Japan, southern India and Malaya hydro-electricity has long been an important power resource. The newer schemes are on a grander scale, embracing whole drainage basins. They are all very costly, and on that account many are unlikely to mature beyond the drawing-board stage for decades. Substantial progress has been made in India where the Damodar river and other schemes are in operation. Of the seven great dams which are planned for the Damodar, four are complete. They enable the Damodar Valley Corporation to control the flow so that more water is available for irrigation and disastrous floods are less frequent. The electricity is used in local industries. Another great scheme is the Bhakra-Nangal on the Sutlej river where one of the dams is over 700 ft. high. The Chinese are working on the Hwang Ho project which will ultimately have over 40 dams. Projects such as these have a tremendous appeal in Asia. Bringing simultaneous benefits to agriculture and industry, they are symbols of economic development and release from poverty. Some Indian politicians have over-estimated their value. The benefits are real, especially in agriculture, but the hydro-electricity will not be anything like enough to make up for the shortage in oil and coal. India, and doubtless other countries, has uranium. It might turn out that nuclear power will become an economic alternative in time to avoid the worst consequences of fuel shortage.

9

India: the British Period and the Partition
of 1947

When the royal charter was granted by Elizabeth I in 1600 to 'The Governor and Company of merchants of London trading into the East Indies', the Portuguese were already well established at Goa and other coastal sites in India. It was hard for the English to gain a foothold. The Portuguese, Dutch and Arab traders were hostile, and powerful at sea, and the Moghul rulers in Agra, who alone had authority to grant trading rights, had a lofty contempt for commerce.

William Hawkins, the first servant of the company to land in India, dropped anchor at Surat in 1608 after a voyage of 16 months. The presents he had brought from England for the Moghul emperor were confiscated by local officials. There were several attempts on his life by the Portuguese, and on his journey to Agra he found that two of his attendants had been hired to kill him. Fortunately, the emperor Jehangir took to him kindly, partly because Hawkins spoke Turkish, but despite the favours of the Moghul court Hawkins was unable to achieve his object: the right to establish trading posts in India. Agreements were drawn up only to be cancelled when the Portuguese made some threat or a tempting offer.

By the time Sir Thomas Roe arrived as ambassador to the Moghul court in 1615, the English had beaten the Portuguese in several local sea fights, and the Moghul attitude had changed accordingly. Roe failed to extract a treaty, but his prestige at Agra protected the English traders on the coast from the rapacity of local officials. The main trading station at Surat prospered. By 1623 there were subsidiary establishments at Broach,

Ahmadabad, Agra and Madras. All these were called *factories* in the language of the company. The *factory* consisted of a warehouse for storing goods (chiefly cotton cloth, spices and dyes), offices, communal living quarters for the traders and barracks for the company's small armed force. In 1647 there were 23 factories and 90 English traders. Pay was low. The traders were expected to make substantial personal profits from the bargains they arranged with Indians. Life was hard, and often dangerous —as on the two occasions when Surat was sacked by Marathas in revolt against the Moghuls and only the factory held out. There was no adequate medicine. Half the English at a factory would die within a few weeks from local diseases. Isolated from home, they adopted Indian dress and customs, but the records suggest they led a temperate life. The English practices of hard drinking and keeping aloof from the local people—not unrelated to the arrival of English women in increasing numbers, especially after the opening of the Suez Canal in 1869—were still unknown.

Such were the small beginnings of British power in India. Great progress was made in the second half of the seventeenth century. The Dutch were still strong competitors, but devoted most of their attention to the islands of South-east Asia. The Portuguese were now in decline. In 1660 their island of Bombay became the property of Charles II, as part of the dowry attached to his marriage with Catherine of Braganza, though the local Portuguese commander did not hand it over until 1665. Charles II, short of money, gave the island to the company in return for a large loan and an annual payment of £10. Bombay was thus the first piece of British Crown territory in India. All the other areas which British traders administered were held by the company by arrangement with Indian sovereign rulers. By 1700 the company's holdings were grouped into the three *presidencies* of Bengal, Madras and Bombay. All were prosperous, but Bengal had great advantages over the others. From Bombay the difficult routes across the steep Western Ghats led only to the country of the hostile Marathas who were to become the strongest indigenous power in India. Madras had access mainly to relatively poor Deccan country. By contrast, Bengal had the wealth of the Ganges lowland and useful river routes penetrating far inland. In the peninsular south the French were strong. For 20 years after 1740 they tried to extend their influence; ultimately they were frustrated by the campaigns of Clive, Stringer Lawrence and Sir Eyre Coote, and their holdings reduced to a few coastal settlements of which the largest was Pondicherry.

During the latter half of the eighteenth century the profits from private trade were enormous. In five years a junior writer in the company could make a substantial fortune. Some returned to live comfortably in England, where their rash spending, strange habits and colourful interpretation of current fashions of dress earned them the nickname *nabob*, a corruption of the Hindustani *nawab* or prince. Many lost their money in the European society of Calcutta, which had become notorious for its debauchery and high living. Life for the English was usually short, for the origins of malaria, dysentery and cholera were not understood. As for the Indians, a third of the population of Bengal was reported to have died in the famine of 1769–70. When Clive returned from his second term as Governor of Bengal in 1767 British authority extended from Calcutta over 800 miles across the Ganges plains almost to Delhi. Bengal and Bihar were directly administered; the rest was ruled by local princes under strong British influence.

An important landmark is the Regulating Act of 1773. By this and later Acts the British government curbed the worst excesses of the company's servants, and created the new post of Governor-General of Fort William (Calcutta) with authority over the other two presidencies as well as Bengal. Calcutta thus became the effective capital of the British in India. Early in the nineteenth century much of south and central India was annexed, to reduce the chaos and destruction caused by warring princes. The last parts of the peninsula to come under British protection were the Maratha states in the hinterland of Bombay. They were annexed in 1818 (over a hundred years after the company administration had been firmly established in Bengal) and Bombay benefited immediately from the wealth of Indian merchants from the Maratha areas. For the first time they were able to go freely to the great port and share in foreign trade. Later they took a prominent part in industrial development.

By 1850 Sind and Punjab had come under British control. The northwest frontier of India was defined at the end of the Second Afghan War in 1880. After the Indian Mutiny of 1857 the British government, as distinct from the company, assumed full authority in India. Already a start had been made on railways and modern irrigation works, and thereafter economic progress quickened.

Two remarkable aspects of British dominion in India are the speed with which it was achieved and the stability it provided for economic development during the following hundred years. The East India Company had sought only trade. Administrative responsibilities over large areas were expensive and therefore to be avoided; they were undertaken only when native authorities were manifestly incapable of maintaining order. The eventual stability of British rule was based on the twofold political division of the territory into British India and the states, made possible by the fact that Indian loyalties were not to one great Indian state but to local leaders and to religion. British India was an area of direct British rule, divided into about 400 districts. It included practically all the best lowland, all the good ports and nearly all the coastline (Figure 11). The states, on the other hand, were independent countries in treaty relationship with the British Crown. The treaties in each instance provided for important matters such as foreign relations to be controlled by the Crown, and for a British Resident to live in the state capital to advise the ruling prince on internal government. Some of the states, notably Hyderabad and Mysore, were very large, with populations of several million and their own armed forces. Others were no larger than English country estates, and were combined for administrative purposes into groups called Agencies. The larger states were separated by British territory, so there was little risk of their combining successfully against the British. But the British themselves were few. In 1861 they numbered 126,000. Of these, 84,000 were soldiers, not many for so large an area only three years after the Mutiny, and 19,000 were women and children. Of the remaining 23,000, most were business men in the towns and ports: only a small minority were district officers, engineers and administrative officers of the central government.

Today India is a country of the Commonwealth and a world power. To countless British of the nineteenth century India meant a career of public service at the end of a three months' voyage and with home leave once every ten years. Before the medical advances of the early 1900's the chances of coming home at all were slight. These people faced great difficulties and tremendous opportunities; viewed from our social surroundings they seem giants of character and determination.

In one of the most fascinating accounts of this time, *India Called Them*,

Figure 11. *India in the late nineteenth century.* The French settlements are numbered as follows: 1—Chandernagore; 2—Yanaon; 3—Pondicherry; 4—Karikal; 5—Mahé

Lord Beveridge describes his parents' life in India for 35 years from the time his father joined the company in 1857. Philip Woodruff's *The Men Who Ruled India* (3 vols.) spans the whole of the British period. The novel *A Passage to India*, by E. M. Forster, is also to be recommended.

ECONOMIC DEVELOPMENT DURING THE BRITISH PERIOD

Substantial progress was made between 1857 and 1947. A good railway system came into being, and by World War II few towns were without some factory industry. The range of industry included textiles, iron, steel and heavy engineering, processing of food crops, leather, chemicals, paper, cement and some non-ferrous metals. The cotton and jute industries demand special consideration because today they employ respectively 28 and 13 per cent of all the factory workers. General and electrical engineering occupies 5 per cent of the factory labour force and the iron and steel industry about 2·7 per cent. While the amount of industry is small in relation to the high population, a great deal of it was the result of local enterprise. Both the cotton and the iron and steel industries were started by Indian business men and almost entirely financed from local private capital. Some 8,000 Indian small investors subscribed to the iron and steel venture. In other fields, especially coal, jute, plantations and railways, British initiative and capital were dominant, though Indian investors contributed.

Railways. Most of the present network was completed by 1900. The broad gauge of 5 ft. 6 in. was chosen for the earliest lines around Bombay and Calcutta, and was intended for the whole system, but shortage of funds later forced the adoption of cheaper metre-gauge lines in many districts. The present mixture of the two gauges shown in Figure 12 is not quite as inconvenient as it looks. All the principal ports, cities and the most productive areas, including the Ganges and Indus plains, are connected by broad-gauge lines. These are mostly of single track, a disadvantage partly offset by the high carrying-capacity associated with broad-gauge rolling stock. The metre-gauge lines serve the less important parts of those same areas, involving transfer of agricultural products at metre-gauge/broad-gauge junctions. There are also four large areas served only by metre-gauge lines: the north-west corner of the Deccan, the country east of the Brahmaputra and south to Chittagong, Kathiawar and much of the southern Deccan. These metre-gauge areas suffer two disadvantages: the carrying-capacity of the narrow trucks is small, and supply of coal from the Damodar coalfields involves inconvenient shovelling from broad-gauge to narrow-gauge trucks. In addition, in the hills there are a few lines of even narrower gauge, such as the Darjeeling railway.

Roads are generally poor by western standards. By 1947 all-weather roads connected the major cities. They are narrow, and often congested with bullock carts which move at an average pace of two miles an hour and rarely keep to the wide earth verges specially provided for them. In the great lowlands lack of hard rock for roadstone hinders the construction of good local roads. Even in the peninsula, where stone is plentiful, the roads are bad. Many are surfaced with lateritic earth. If rolled flat at the end of the wet season this material quickly sets into a hard smooth surface. In the dry season it becomes increasingly dusty, and by the next monsoon there is a thick layer of powdered laterite which the rain converts

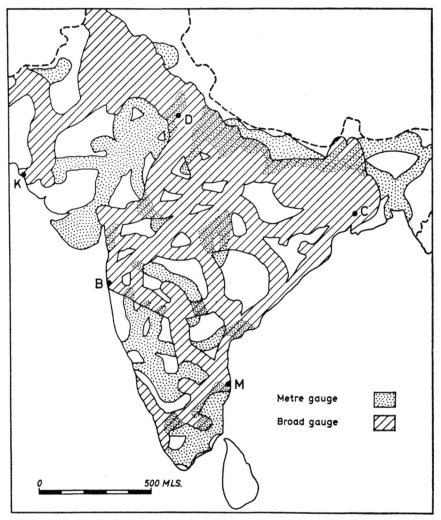

Metre gauge

Broad gauge

0   500 MLS.

Figure 12.   *India: railways* Land within 50 miles of a broad-gauge railway is cross shaded. Land within 50 miles of a narrow-gauge line is stippled

into deep mud. During the monsoon the going is bad even for carts, and impossible for motors, especially in paddy areas where the unfenced roads run on embankments 10 ft. or more above the flooded fields. Vehicles which skid into ricefields stay there until the next dry season. Thousands of villages are completely isolated during the summer monsoon.

River transport decreased in importance on the rain- and snow-fed Indus and Ganges systems as those rivers became increasingly used for irrigation. The peninsular rivers, dependent on the summer monsoon and therefore with little water in them for half the year, were never worth developing as means of transport. Steamer services became vital economic links only in the Ganges-Brahmaputra delta and along the Brahmaputra into Assam.

The cotton industry (Figure 13). In the time of the East India Company there was a thriving export of fine hand-made cotton goods to Europe. Muslins from Dacca in Bengal were famous, and cloths from southern India were named calicoes after the port of Calicut. British legislation to protect the Lancashire cotton industry brought a decline in this commerce during the eighteenth century, while cotton goods from Lancashire mills sold in India at prices well below those of the local products. The Indian handicraft industry almost died out, and when a Parsee merchant started the first cotton mill in Bombay in 1851 there was no local textile tradition on which to base the modern industry.

Bombay rapidly became the principal cotton-manufacturing centre of India. At the outset a coastal location was an advantage because there was no railway to bring fuel from the Damodar coalfields. Coal, machinery and other factory equipment were imported from England. The Bombay mills produced only cotton yarn, which was chiefly exported to China where it almost monopolized the market. Madras and Calcutta would have been equally suitable sites for the industry, if access to the Chinese market and supplies from England had been the only factors involved. Cotton is widely grown in India: the largest production in those early days was in the western Deccan, but later, with the spread of irrigation in the Punjab, the principal cotton-growing area embraced the large tract shown in Figure 13. Bombay clearly had some advantage over the other two cities as regards proximity to that area. More important, however, was that Indians with capital and business ability

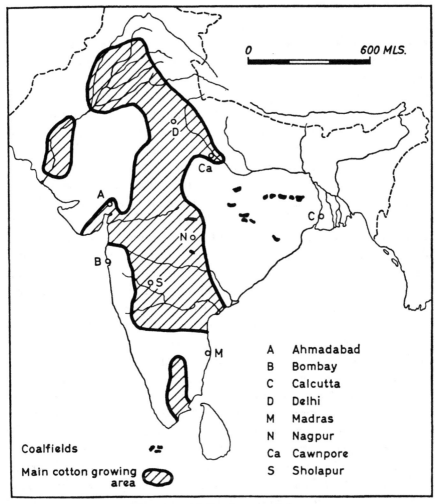

Figure 13. *India: the cotton industry*

sufficient to start the industry were to be found only in Bombay and
neighbouring west coast ports where they had prospered for generations
on Indian Ocean trade, and in the Maratha country which had been free
of British domination until 1818. Elsewhere, and especially in Bengal and
Madras, the British had monopolized commerce for so long that Indian
initiative was not forthcoming. The existence of Indian capital and enterprise in the Bombay area alone was the most powerful single influence in
the choice of that city rather than Calcutta or Madras.

At the turn of the century important changes reduced the advantages of Bombay as a centre of cotton manufacture. Japan became a strong competitor in the Chinese market, and even invaded the internal market of India. By 1925 over three quarters of the Chinese trade was in Japanese hands, and the Indian industry had to adjust itself to supply the internal market of India. By that stage there were railways to carry Damodar coal throughout the land, and cotton mills sprang up in increasing numbers wherever the local market was attractive. Also, the Indian market was for cloth, not yarn, so the weaving industry, hitherto retarded, developed quickly. In 1900 Bombay's 90 mills had represented half the total for the entire country. By 1947 there were over 400 mills, many of them in provincial towns, where labour, food and building land were cheaper than in the cities, but the crowded island of Bombay still remained supreme as the centre of the industry. Of the other mill towns, only Ahmadabad, Sholapur, Madras, Calcutta, Cawnpore, Delhi and Nagpur served markets beyond their immediate neighbourhoods. The provincial mills produced mainly coarser cloths, while Bombay and Ahmadabad specialized, as they still do, in finer fabrics based in part on imported Egyptian cotton.

Jute is an annual crop that grows exceptionally well on land inundated each year by silt-laden flood water in the Ganges-Brahmaputra delta and lowland Assam. An exhausting crop, it requires costly manuring when grown on higher ground. There was a traditional local handicraft industry which converted the fibre into coarse material for sacks, screens and cordage. The modern factory industry first developed at Dundee in the early nineteenth century in response to the world demand for sacks for transport of grain, cotton, wool and other products. The first Indian jute mill was opened in 1855 near Calcutta. By 1947 there were about 100 mills, almost all of them along the banks of the Hooghly within 30 miles of Calcutta. This Indian industry flourished on oversea trade in sacks and sacking, but when sacks became less in demand more attention was given to tarpaulin sheets and tenting. Until recently the industry was almost exclusively British in finance and management—most of the mill managers being Scottish. British business houses in Bengal had the advantage over possible Indian promoters in that they had oversea connections which could secure large orders in Australia and America. There is also evidence that British banks in Calcutta discriminated against Indians in arrangements for credit.

The concentration of the industry along the Hooghly is a striking illustration of the influence of physical geography and communications on industrial location. The mills needed raw jute, coal from the Damodar area and access to an ocean port. With those requirements in mind, one might have expected the mills to be sited in the jute-growing area or on the coalfield, or at Calcutta, or even throughout Bengal. But the operative consideration was that the conveyance of fuel or raw material beyond Calcutta involved expensive handling. Coal reached Calcutta easily by broad-gauge railway. To carry it eastward into the jute-growing country —where most of the railways are of metre gauge and there are no bridges over major waterways—meant trans-shipment from railway wagons into river boats. In the opposite direction, raw jute was easily collected by thousands of small river boats and brought to Calcutta along the maze of delta channels. Carrying jute westward beyond Calcutta involved transfer to the railway.

Iron and steel. The best iron-ore deposits in India are shown in Figure 14. These, with iron contents between 60 per cent and 70 per cent, are so rich that the enormous but low-grade deposits elsewhere in the peninsula are unlikely ever to be needed. This wealth was first recognized when geologists working for the Indian industrialist Mr J. N. Tata discovered near Drug a hill of solid iron ore 300 ft. high and with an iron content of nearly 70 per cent. Even richer deposits were proved in the Mayurbhanj district, and these became the principal source for the Tata plant, which began production in 1911 at a site near the Subaranekha river. Formerly an insignificant village, this place became the industrial town of Jamshedpur. By 1951 the population was 220,000. The coking coal came from the Jharia field, which alone among the Indian coalfields has substantial amounts of coal of suitable quality. It is reasonable to expect that an iron and steel industry might have started also at that coalfield to which returning coal trains could have brought iron ore. Such a development did occur, but it was at Asansol on the Raniganj coalfield, 30 miles farther east, where there was not much coking coal. This was related to the better water supply at Asansol, where the Damodar is reinforced by its main tributary, the Barakar.

Another interesting though small-scale works grew up in the south of the peninsula at Bhadravati, where local ore is smelted in charcoal furnaces and the steel plant uses hydro-electricity. The charcoal comes from surrounding forests which are carefully cut and replanted.

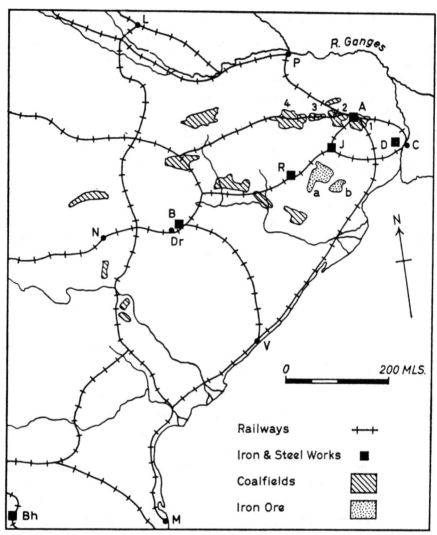

Figure 14. *India: the iron and steel industry* The four principal coalfields which together account for 92 per cent of the Indian reserves of all coal other than lignite, and for practically all the reserves of coking coal, are numbered as follows: 1—Raniganj; 2—Jharia; 3—Bokaro; 4—North Karanpura. Of the reserves of coking coal, Jharia has 60 per cent, and Raniganj and Bokaro each have about 20 per cent

The iron and steel centres are indicated thus: A—Asansol; B—Bhilai; Bh—Bhadravati; D—Durgapur; J—Jamshedpur; R—Rourkela. Iron ore fields are shown as follows: a—Singhbhum; b—Mayurbhanj; Dr—Drug. Other towns indicated are : C—Calcutta; L—Lucknow; M—Madras; N—Nagpur; P—Patna; V—Vizagapatam

5. Foothills of the eastern Himalayas: a farmstead and a flight of paddy terraces on a steep hillside

6. December scene on the Bengal plain, near Comilla. Gaunt cattle graze rice stubble in the tiny fields. In

Before World War II India produced pig iron at lower cost than any other country, and even sold iron on the west coast of U.S.A. at competitive prices. The steel industry was less flourishing and needed protection from the competition of imported steel. Indian iron and steel products covered only a few of the items required by the local market. Cast-iron sleepers, tin plate, steel plates, bars and rails were made, but all alloy steels were imported.

Irrigation. Probably the greatest and most permanent achievement of the British in India was the provision of modern irrigation works. The largest scheme, fed from the mile-long Sukkur Barrage on the Indus in Sind, has over 36,000 miles of distributary channels and waters nearly as much land as the total cultivated area of Egypt. Yet the Sukkur scheme alone is today only one sixth of the area irrigated by modern canals. Other great schemes are those of the Punjab and the Ganges plains, and the Kistna, Godavari and Cauvery deltas of the peninsula. All these works provide perennial irrigation in which the supply of water is carefully regulated. They are a tremendous advance on the inundation canals of earlier centuries which provided water only when rivers were in flood, and then only to land adjacent to the rivers. The modern canals are long and some carry water to farmland over 100 miles from the rivers. The Sukkur and many of the Punjab works brought water to arid and largely uninhabited country: new compact farms of economic size were created without the hindrance of rearranging an established rural population with small and fragmented holdings. An area of 25 acres is a common size and some farms exceed 50 acres.

A later development are the tube wells of the upper Ganges plains. These deep wells tap the great reservoirs of water deep in the alluvium. Many are worked by hydro-electricity generated on the river-fed irrigation canals where these drop in level by 10 ft. or so. Figure 25 (page 133) shows some detail of the Punjab canals. In most of peninsular India agriculture is dependent on tanks and shallow wells which are less efficient than the modern canal systems.

Plantation agriculture. Prosperous tea estates multiplied in Assam, the Himalayan foothills of north Bengal and on a smaller scale in the Nilgiri Hills of south India. Economically far less important were small coffee estates in Mysore and rubber estates in the coastlands of Travancore.

The last 50 years of British rule were marked by the rapid growth of national consciousness among the Indian people. There were frequent riots and widespread impatience for independent status, though much bloodshed was avoided thanks to the non-violent policy of the principal Hindu leader, Mahatma Gandhi. The main problem was the form independence should take. The largest political party, the Indian National Congress, excluded no one from membership on grounds of race, religion or language. But three quarters of the population were Hindu, and democratic government of India as one state could mean only government by Hindus, whatever provision was made for the representation of the remaining quarter, who were mainly Muslims. The fears of the Muslim minority were justified on grounds other than mere numbers, for the growing Hindu middle classes had embraced western education. Many Hindus had been educated in Europe and America, and the Indian universities were a growing force. The Muslims on the whole had remained aloof from these trends. They were distrustful of British good intentions. Many felt strong cultural ties with the Islamic world of the Middle East and were offended by the British participation in the abolition of the Ottoman empire in 1918. Suspicious of Hindu aims, and led in the later stages by the western-educated Mr Muhammad Ali Jinnah, the Muslims campaigned for their own national state of Pakistan.

British efforts to secure independence for an undivided India were maintained to the last. In 1947 partition into two states, India and Pakistan, was accepted as the only practicable solution. Pakistan was to consist of the two areas where Muslims were a majority—in the Indus lowland and in Bengal—but the precise boundaries between India and Pakistan presented a staggering problem. Both in the Punjab and Bengal there were border zones where Muslims and Hindus lived side by side in substantial numbers. In the Punjab there was the additional complication of the Sikhs, a fierce and proud minority of six million among the Muslims, with a religion akin to Hinduism. The boundary commission based its award on the population data for small administrative divisions, with the result that parts of old local boundaries, not unlike those of English parishes, were linked to form international borders. Communal and religious feelings were so strong that no other solution would have served, but the consequences were costly. Large numbers of Muslims, Sikhs and Hindus found themselves on the wrong sides of the boundaries. About 10 million people left all their fixed possessions and fled: Muslims into Pakistan, Hindus into the new India. Many were cut to pieces on the roads by their countrymen. Others died in their burning homes. Riots

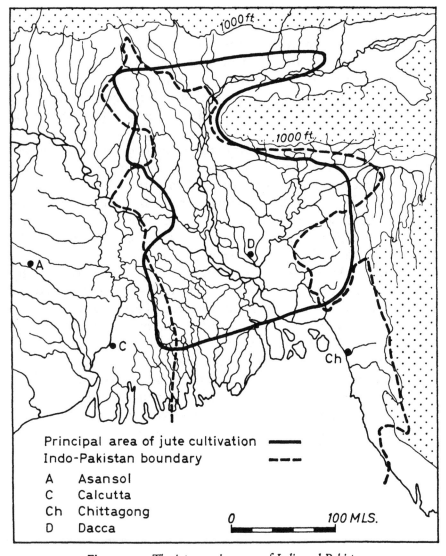

Figure 15.   *The jute-growing areas of India and Pakistan*

and killings were common in the British period, but savagery on this scale had been unknown since the heyday of the Marathas. Minorities elsewhere in the country were more fortunate. The Buddhists—perhaps 250,000 of them, mainly in the foothills of the Himalayas—were not materially involved. The 100,000 rich and prosperous Parsees in and around Bombay had a secure place in India.

The new states inherited the resources of the old undivided India in the following proportions. Pakistan, with only a fifth of the total population, included half the irrigated land and an even larger share of the area watered by modern canals. The new India, with four fifths of the population, included all the heavy industry, all the iron ore and other minerals of the peninsula, practically all the coal and all the good ports except Karachi. The details of the new boundaries, harmless enough in their former status of local administrative limits, brought economic confusion and loss. In the west the boundary crossed and re-crossed main railway lines; it ran down the centre of single-track lines and cut through the irrigation schemes shown in Figure 25. Canals from barrages in India flowed across to water fields in Pakistan. The Indians, short of good irrigation schemes, were accused of reducing the flow across the border. This dispute took 13 years to settle. Figure 15 shows how much of the jute-growing area of Bengal was separated from the milling area, which was entirely in India: the effects of this have only partly been offset by the increase in the Indian jute acreage. The India-West Pakistan border remains a source of tension, with military forces of both sides at the ready. No agreement has been reached on Kashmir, where Indian and Pakistani forces face each other across a temporary cease-fire line.

Confusion sometimes arises over the two meanings of the word India: the India of pre-1947, which occupied the entire sub-continent, and the modern political unit which was created in 1947. In all further references to India in this book the second of the two meanings is intended, except where the context clearly embraces the pre-1947 period.

10

The Republic of India

ECONOMIC AIMS AND ACHIEVEMENTS

Shorn of the irrigated lands of the western Punjab and the Indus, which normally produce food grains surplus to local needs, the new republic of India faced at the outset a critical food problem. There were massive imports of food, but that was not acceptable as a long-term solution because it drained the small reserves of foreign exchange which were needed urgently to buy capital goods for industrialization. Imports of food continue, but the gap between production and the amount of grain required to maintain the minimum diet has been progressively shortened. This has been achieved mainly by the extension of irrigation, though better manuring and seed selection and even some consolidation of holdings have helped. At the beginning of the first five-year development plan in 1951 about 51 million acres, or one sixth of the net sown area, were irrigated. By the end of the plan in 1956 the irrigated area had risen to 66 million acres. By 1981 it might reach 120 million acres: together with more manure and better seed this would allow much higher yields per acre, but whether it can solve the food problem depends among other things on the population increase. In 1961 the population was 442 million. The highest estimate for 1981 is 682 million, and the lowest 529 million. One difficulty is that most of the river sites which could be cheaply exploited for irrigation have already been developed. The barrages of the Punjab schemes were easy to construct because they were in the plains. Only a low retaining wall was necessary to impound a large volume of water. Today the engineers have to look for sites higher upstream in hilly country, which demands expensive high dams such as those on the Damodar and on the Bhakra-Nangal scheme of the Sutlej. Provision of a balanced diet, including milk and meat, would require a lot of land for fodder crops and consequent higher yields from fields remaining under grain.

While Indian capital financed a good deal of industry during the British period, there is not enough of it to secure the rate of progress needed today. Three new iron and steel plants have been built with technical and financial assistance from abroad: at Rourkela with German help, at Bhilai with Russian help and at Durgapur with British help (Figure 14). These three plants, together with improvements at Jamshedpur, Asansol and Bhadravati, will increase Indian steel production to six million tons by 1963. There is plenty of manganese ore and chrome for use in special steels. Foreign capital is essential: some comes from U.S.A. and U.S.S.R., though in 1960 British capital amounted to 80 per cent of the total foreign investment in India. The leading companies in most branches of British industry are co-operating with Indian firms in new developments ranging from electrification of railways, heavy engineering, electrical, machine-tool and aero-engine manufacture to plastics, paints and antibiotics. But methods devised in western countries, where labour is scarce and expensive, require adjustment before they can be fitted into the present Indian economy. In India labour is cheap, and the labour force is expected to grow at least until the mid 1970's, so industrial techniques that are *labour-intensive* pay off better socially and economically than methods which employ few people. It is to be expected that as the Indian people become more prosperous the value of human labour will rise relative to that of capital equipment, and labour-saving techniques will become more economic. Such a rise in labour cost will be beneficial to farming if it attracts people away from the land into industry. It would make consolidation of holdings and mechanization easier.

The multi-purpose hydro-electric schemes were begun after 1947 and are therefore proudly associated with independence. The principal works now completed or in hand, and the benefits expected from them, are listed in Table 5. The ultimate generating capacity of the works is a mere two million kilowatts. Some authorities rate the hydro-electric potential of the whole country as high as 40 million kilowatts, and they estimate that the installed capacity of the hydro-electric stations will be 17 million kilowatts by 1975. By 1981 the installed capacity from water power and thermal sources is expected to be 35 million kilowatts, a tenfold increase on the 1958 figure.

These developments in agriculture and industry are being attempted by a democratic state with a 'mixed' economy in which some industries are nationalized while others are private enterprises. Similar problems are being tackled in China under a Communist dictatorship in which the administration can convert all property to public use, abolish individual peasant holdings in favour of large collective farms and direct labour to whatever tasks are thought most vital. The political future throughout Monsoon Asia depends largely on which of these two great states is the more successful in solving its economic problems. Failure in India could mean the end of democratic freedom there and wherever else it now exists

TABLE 5

INDIA: BENEFITS EXPECTED FROM HYDRO-ELECTRIC PROJECTS

Project	Location	Irrigated acreage	Generating capacity (kw.)
Bhakra-Nangal	R. Sutlej	3,424,000	517,000
Damodar	R. Damodar	1,326,000	274,000
Hirakud	R. Mahanadi	1,930,000	268,000
Tungabhadra	on tributary of R. Kistna	1,000,000	60,000
Numerous smaller schemes	—	9,219,000	845,000
		16,899,000	1,964,000

in the smaller states. Some think that in the present stage of Indian development democratic processes are too expensive a luxury, considering the time it takes to get things done, and that for a limited period some kind of authoritative régime will be necessary if an acceptable rate of progress is to be maintained. Others, impressed by the rapid increase in numbers, speak of an 'explosion' of population against which both the Chinese and Indian attempts will be futile.

While food production and industrialization are today urgent problems demanding all the direct foreign help that can be given, it is interesting to speculate on other potentialities that might be realized in the future. Among those, tourism, which has brought economic benefits to many countries, is surely outstanding. When long-distance air travel becomes cheaper, the mountains, cities and beaches of India will be easily accessible from Europe.

In his book *India and Pakistan* Professor O. H. K. Spate recognizes 35 main regional divisions in the sub-continent and 75 divisions of a second order; these in turn are broken down into about 225 subdivisions. Most of these divisions are recognizable on grounds of topography or culture or both, but some 'are simply what is left over when more definite units have been sieved out'. Spate emphasizes the need for detailed local studies by Indian geographers, which alone can provide the basis for sound regional analysis. These facts are stated lest the following paragraphs, which convey only the principal regional contrasts of the country, should give a deceptive impression.

(1) THE NORTHERN LOWLANDS

This is an enormous alluvial area, 1,000 miles from east to west and about 200 miles wide, bordered in the north by the sharp line of the Siwalik Hills and the jungly foothills of the Himalayas, and on the west by the Indo-Pakistan boundary. To the south there are broad alluvial salients into the peninsular uplands along the valleys of the Chambal, Betwa, Ken and Son. There is no sharp physical or cultural change to serve as an eastern boundary, but the border of West Bengal, where the Ganges changes to a southerly course, is a convenient limit. This eastern part is a well-watered area to which abundant trees scattered throughout the cultivated land give a green appearance even in the dry season. Westward, rainfall decreases, trees are smaller and fewer, except along the northern border, and everywhere is brown and parched in the dry season. Beyond Allahabad agriculture becomes increasingly dependent on irrigation. In the south-west the region merges with the Thar Desert. Densities of rural population reach 1,000 per square mile over large areas in the east, about 700 in the plains around Delhi, but are much lower on the arid south-western borders. There are several major towns, among them Amritsar, Ambala and Meerut, marking the main route into the Ganges lowland from the north-west. Others, such as Patna and Lucknow, are old regional capitals, while Allahabad and Banares are places of pilgrimage for devout Hindus. But none of these has locational advantages to compare with

Delhi and Agra. These two cities command a key position from which successive invaders have controlled all the plains and much of the peninsula; and they dominate not only the routes along the northern lowland but also the routes into the peninsula and to the ports of Kathiawar.

Delhi. Alone among the towns of the plains, Delhi is built on hard rock which rises from beneath the alluvium. This rock forms the Delhi ridge, a northward extension of the Aravallis which projects in places some 400 ft. above the neighbouring plain.

Some of the striking facts about Delhi are indicated in Figure 16. The ridge is an ideal situation from which to control the Jumna crossing and the plains to the north. The city known today as Old Delhi (otherwise Shahjahanabad) was built by the Moghul emperor Shahjahan who moved his court there from Agra in 1648. The walls of Old Delhi, shown as a thick black line, are 30 ft. high. Inside is the Red Fort, built of red sandstone, and a maze of narrow streets and alleys shown by the pattern of thinner lines. When the British first settled in Delhi they built to the north of the Moghul city, where the ridge comes very near the river, in the area indicated by the more open pattern of roads. In 1911, when it was decided to move the capital of British India from Calcutta to Delhi, work began on a new city, New Delhi. There are magnificent public buildings well arranged on broad avenues, with room for expansion to the south.

In terms of Indian history even Old Delhi is recent. At least eleven earlier cities were built in association with the ridge—the earliest known only from legend. Some were sacked in war. Each in turn was plundered for bricks and stone to build a successor. Today substantial remains of foundations and defences of three can be seen: Rai Pithora (twelfth century), Siri (early fourteenth century) and Tughlaqabad, begun in 1321.

The rivers. The Beas and Sutlej rise in the Himalayas and flow southwest across the Pakistan border to join the Indus. The Ganges and its major tributaries (Jumna, Gogra, Gandak and Kosi) also have Himalayan catchments, thus deriving water from melting snow as well as from rain. The flow of the Son and Chambal, the main Ganges tributaries with peninsular catchments, is entirely dependent on summer monsoon rain.

The Siwalik range, which rises to 5,000 ft. between the Beas and the Jumna, consists of very recent gravels, clays and unconsolidated sands. The hills were once forested. Severe gully erosion followed extensive

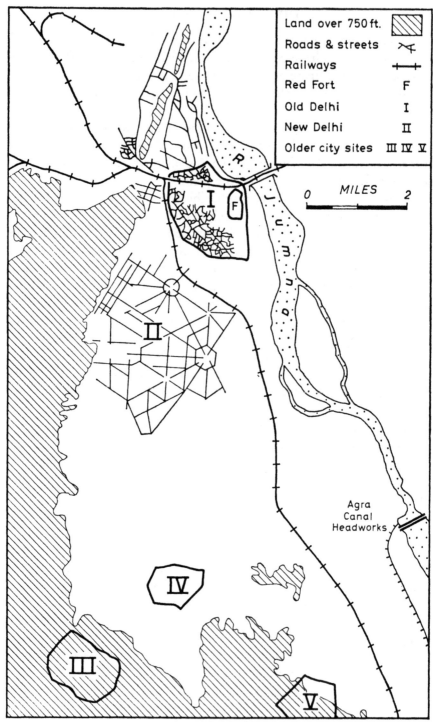

Figure 16. *Five of the thirteen cities of Delhi* The three older city sites are
III—Rai Pithora; IV—Siri; V—Tughlaqabad

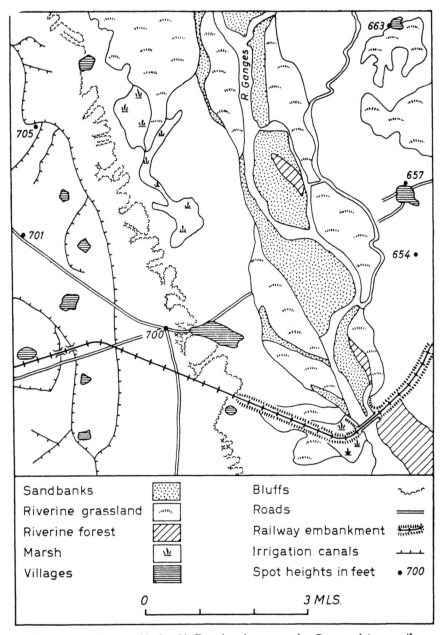

Sandbanks		Bluffs	
Riverine grassland		Roads	
Riverine forest		Railway embankment	
Marsh		Irrigation canals	
Villages		Spot heights in feet	• 700

0   3 MLS.

Figure 17. *Bhangar, khadar, bluffs and settlement on the Ganges plain 30 miles south-east of Meerut* The spot heights indicate that the bluffs separating the *bhangar* and *khadar* are here about 50 ft. high. Only the *bhangar* has canal irrigation. The railway crosses the *khadar* on an embankment. Throughout the plains of India and Pakistan embankments of this kind interfere with natural drainage and increase the flood hazard in districts on the upstream side

clearing of the forests during the last 150 years. The torrent courses, known locally as *chos*, are dry except during the rains, when vast loads of sand are swept southward. On the plains the *chos* frequently change their courses. Hundreds of acres of good farmland are thus covered by gravel and sand spreads, and are lost to cultivation (Figure 20). Some of the *chos* are being brought under control by the planting of grasses and shrubs on their banks. Few have enough water to persist far into the plain, but some unite to contribute a seasonal flow to the Ghaggar (or Sarasvati). Indian legend tells of a great river, the Sarasvati, flowing through rich agricultural lowland: in the wet season water usually flows along the Ghaggar channel for 300 miles. At one time there was certainly at least a seasonal flow in the dry section of the Ghaggar, which can be traced for a further 200 miles in arid country on a course parallel to and south of the Indus (Figure 25). Ruins of about 40 towns and villages closely associated with this dry course indicate a reliable flow during the time of the Indus civilization. As to the causes of this deterioration of the Ghaggar, current opinion suggests that destruction of forest and the spread of agriculture during 3,000 years on the plains had far-reaching effects on ground water. Natural or artificial diversion of former headstreams into the Sutlej or Jumna could have been an influence, while in recent times the building of road and railway embankments at right angles to the general slope of the plain has certainly interfered with natural drainage, probably to the disadvantage of the Ghaggar.

The rivers of the northern plains have migrated to and fro over the alluvial tract. Where the courses are approximately north-south there is a tendency, on account of the rotation of the earth, for rivers to cut into their right banks. Right banks are therefore generally higher than left banks, and it is easier to take off irrigation channels from the left banks. Two divisions of the alluvium are recognized, the older alluvium (*bhangar*) of Pleistocene age and the newer alluvium (*khadar*). Usually the *bhangar* occupies the higher levels of the *doabs* (the land between the rivers), while the *khadar* forms the flood plains. The *doabs* are broad flats rising 100 ft. or more above the rivers. The edge of the flood plain is marked by bluffs sometimes 100 ft. high. Formed in alluvium, these are easily eroded. Along the Chambal and Jumna they are so severely gullied as to form badland tracts in places several miles wide. Elsewhere the bluffs form good settlement sites above flood water (Figure 17).

The northern lowlands have three main divisions: *the Indo-Gangetic divide, the upper Ganges plains* and *the middle Ganges plains.* The division of the Ganges plains into upper and middle sections is necessary on account of climatic and agricultural differences, but the transitions are gradual and the selection of a dividing line is a problem. Drawing attention to the absence of a sharp physical break, and to the fact that the isohyets trend not north-south but N.W./S.E., cutting the rivers at acute angles, Spate favours the demarcation suggested by Stamp, the line joining Allahabad with the N.N.W./S.S.E. section of the Gogra (Figure 18). Thus defined,

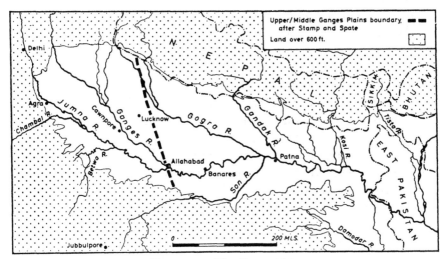

Figure 18. *The Ganges plains*

the upper Ganges plains to the west are a zone of mixed crops and include most of the irrigated land of the Ganges lowland, while in the middle Ganges plains to the east rice is dominant. But no sharp change occurs at the line. Canal and tube well are the main systems of irrigation, except along the peninsula fringes where there are tanks.

The Indo-Gangetic divide

This is the country between the Jumna and Sutlej, and it is convenient to include also the area westward to the Pakistan border, i.e. the Bist-Jullundur *doab* and the plains around Amritsar. The northern edge, within 50 miles of the Siwaliks, has an appreciable winter rain in addition to the summer monsoon. There the water table is high and wells are easily constructed. Well irrigation supports wheat, maize and sugar cane, except where the land has been devastated by *chos*. Winter wheat, watered by canals, is also the main crop in the drier country to the south. Rain-fed crops are gram and barley (*rabi* crops) and millets (*kharif* crops)—*rabi* and *kharif* being local terms for winter and summer crops respectively. *Rabi* crops are harvested about March. *Kharif* crops grow during the summer monsoon rain and are harvested in autumn or even later. It is important to remember that crops such as sugar cane, which occupy the ground for more than a year, and also some rice crops, do not fit into this simple classification.

In the southern borders, where the average rainfall is below 12 in. and the water table very deep, cattle and camels are more important than cultivation, except where canals from the Sutlej, Ghaggar and Jumna bring additional water. More of this arid country will become productive as canal irrigation is extended; much of the water will come from the Sutlej on completion of the Bhakra-Nangal scheme (Figure 19).

Delhi has a population of over one million, but of the other towns only Amritsar exceeds 250,000. All are market centres and have some industry, mostly textiles and food-processing. Jullundur and Ludhiana are also railway junctions. Ambala rose to importance as a military station in the British period. Chandigarh, the new East Punjab capital, is near the hills on the railway north of Ambala (Figure 25).

The upper Ganges plains

Wheat (*rabi*) and millets, rice and maize (*kharif*) are the grain crops, with some barley (*rabi*). Winter rain in the extreme north-west and along the Himalayan border as far as the Gogra is enough for wheat to do well without irrigation, but south and south-eastward that rain fades out and wheat becomes increasingly dependent on canals and tube wells, despite the greater annual rainfall towards the south-east. The opposite is true of rice, which increases in importance eastward as the summer monsoon rain improves. In the north-west the small rice acreage is irrigated, because the summer rainfall is low. Towards the east almost the only irrigated rice-fields are those naturally flooded in the *khadars*. But most of the ricefields are on the *bhangar*: they are also under water while the rice is growing, but that water is simply the accumulation of rain falling within the fields. Sugar is a cash crop, principally along the northern, wetter edge of the plain. Other cash crops are oilseeds and cotton, the latter less prominent today than formerly.

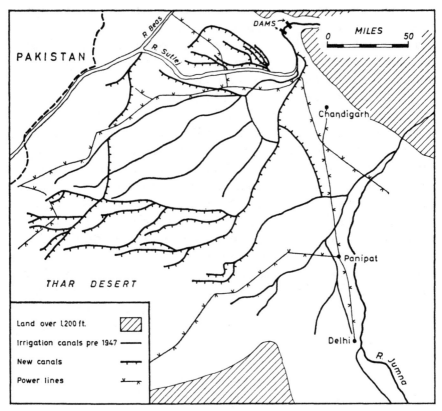

Figure 19. *The Bhakra-Nangal scheme* Note the location of the dams, the new power lines, and the new canals which bring water to the arid borders of the Thar

Cattle, far too numerous, graze in the stubble after harvest. When the fields are under crops they make do with whatever they can get on roadsides and the few patches of wasteland. By the end of the dry season they are too weak to work. Equally, there are far too many people. Density of rural population reaches 1,000 per square mile over large areas. The average size of holding is below five acres. Here, as in most of India, any plan for agricultural improvement which involves taking land out of cultivation for a few seasons, and so temporarily reducing food production, is unacceptable because the majority of the farmers currently need all the grain they can raise. Few can build reserves of grain and money to tide themselves over until improvements such as new rotations begin to yield. Only under grain can the land produce enough to keep the population alive. If a substantial area were given to fodder crops to support better cattle, which in turn would yield meat and dairy products and so contribute to a better diet, the *total* food production per head of population and per acre would be less. Any major dietary improvement must await a reduction in the population.

Of the major towns, Agra (population 462,000) on the Jumna is one of the most fascinating. Though squalid, like practically all Indian towns, it was one of the Moghul capitals. The Taj Mahal (tomb of the emperor Shahjahan's favourite wife), the Pearl Mosque and the fort display Moghul architecture at its best. Twenty-five miles west of Agra the emperor Akbar built another capital, Fatehpur Sikri. Begun in 1570 and abandoned about 40 years later, this walled city, with mosques, palaces and impressive red-sandstone gateway, is a miracle of Muslim art. One of the courts is laid out in squares of red sandstone: there the great Akbar played chess, with slave girls as chessmen. Allahabad (population 413,000) is a legal and administrative centre with some industry. Lucknow (population 600,000) has railway workshops, paper and cotton mills, and sugar refineries. Cawnpore rose to prominence during the American Civil War. Newly connected with Calcutta by railway, it was a good collecting point for Indian cotton to be sent to markets normally supplied from America. Cawnpore now has wool, leather, cotton, flour and chemical works. At the town of Hardwar on the northern edge of the plains are the headworks of the Upper Ganges Canal (Figure 20).

The middle Ganges plains

Moving eastward into the middle Ganges plains the proportion of *khadar* increases at the expense of the *bhangar* as the flood plains widen. Rainfall becomes greater. At the eastern end, where the annual rainfall is about 70 in., there is hardly any *bhangar* except at the Himalayan and peninsular edges. The summer monsoon, which arrives earlier than in the Delhi country, and pre-monsoon showers in May, give a longer wet season. The simple division of crops into *rabi* and *kharif* applies well enough on the western border where wheat is still prominent, but is progressively replaced eastward by a system in which crops are planted so as to achieve the following three distinct harvests:

1—the *bhadai* (or autumn) harvest—maize, millet, quick-growing rice. Also some jute on eastern margins;

7. Rourkela: one of the newer iron and steel plants of India

(Photo—Press Information Bureau, Government of India)

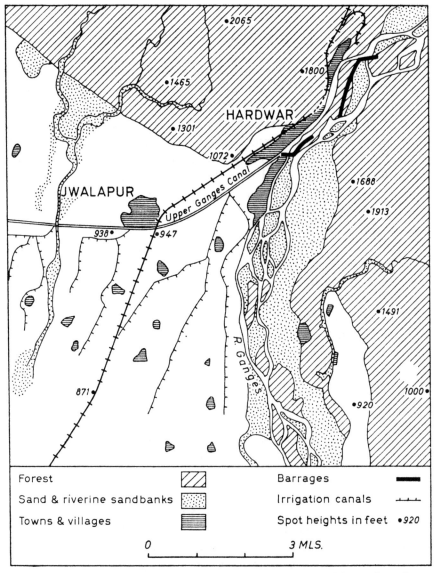

Figure 20. *The Ganges Canal headworks* The sharp break of slope where the plains meet the Siwalik Hills is marked by the straight-line boundary of the forested area in the north-west where three *chos* have united and brought spreads of sand to the plain. The Ganges emerges as a braided stream from the Siwalik Hills at Hardwar, and flows southward. Two long barrages at Hardwar suffice to divert water into the Upper Ganges Canal, whence several irrigation channels distribute the water southward. The spot heights indicate the relief of the plains and hills

2—the *aghani* (or winter) harvest—the main rice crop;

3—the *rabi* (or spring) harvest—catch crops, gram.

The *aghani* rice crop, which grows throughout the wet season and is harvested in December and January, is the main food supply. Obviously, all three harvests cannot be had from a single piece of land in one year. Most of the land is given to the main rice crop, and is planted with *rabi* crops after the *aghani* harvest. Generally a smaller portion of a farm is planted with *bhadai* crops, to be followed by *rabi*. But the *rabi* crops here are of little value compared with the *rabi* wheat and barley which are principal food crops of the drier country with cooler winters farther west. The monsoon rain dies away in October, and drought in that month can damage *aghani* rice and give poor conditions for starting the *rabi* crops. All this section of the northern lowland is frost-free. The only extensive canal irrigation is in the Son lowland, though there is tank irrigation on the southern border of the plain. Rural densities of population exceed 1,000 per square mile over large areas.

There are three cities. Banares (population 356,000), one of the holy places of Hinduism, is always thronged with pilgrims who come to bathe in the sacred Ganges. Patna (population 364,000) is an administrative centre built on the site of the ancient Pataliputra. Gaya is a place of Buddhist pilgrimage. All three, of course, have some industry.

The rivers are often a menace. A good deal of riverine land is left to a natural vegetation of grass and forest. Large areas are occupied by abandoned channels which, when full of flood water, form lakes often 20 miles long. North-bank tributaries with catchments in the Himalayas are a special danger; they scatter sand and gravel over the northern part of the plain and the water discharge is enormous. The Kosi, with a large catchment including Mount Everest, has built an inland delta of sand 50 miles across. There is a plan to build two great dams on this river, one in Nepal with a height of 750 ft., the other 30 miles downstream in India, which would then yield irrigation water and power for both countries.

(2) THE UPLANDS OF CENTRAL INDIA AND THE PENINSULA

This area extends from the Thar Desert and Aravalli Hills of the north for 1,200 miles southward to the Cardamom Hills, and 1,000 miles south-eastward towards Calcutta. Though plateau country predominates, there are also scarplands, wide river valleys and the narrow faulted trenches of the Narbada and Tapti rivers. There is much variety in physical geography and in the cultures: parts have been studied in detail by Indian geographers, but a complete regional analysis must await the extension of their work to the whole. The following paragraphs indicate merely the more striking regional subdivisions which are shown in Figure 21.

The name Deccan is generally applied to the more strictly peninsular part, i.e. the country south of the Satpura Hills, but some writers use it to

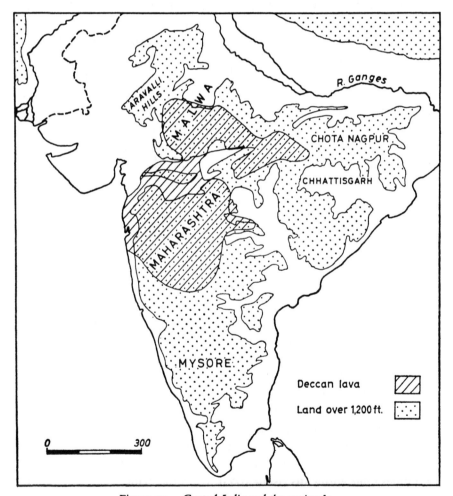

Figure 21.   *Central India and the peninsula*

denote the whole area. Almost a quarter of the country, in the west and north, is covered by the Deccan lavas (Figure 21). These beds, for the most part nearly horizontal, impart a mesa-and-butte topography to large tracts. The heavy black soils (*regur*) associated with them are moisture-retaining, an agricultural advantage where the rainfall is low and unreliable. The best *regur* has accumulated in valley bottoms; thinner but still good *regur* covers the higher ground. The upland soils in the parts not covered by lava are generally lighter and poorer than *regur*, but there are useful alluvial soils in the river basins of the Mahanadi, Godavari, Kistna and Cauvery.

In the north the Aravalli ridges rise from 1,000 ft. immediately south of Delhi to 5,500 ft. in the south-west. The ridges are bordered by plateau at a general level of 1,500 ft. While the higher Aravalli slopes in the south get over 40 in. rainfall, the lower ground is arid; the *nullahs* (torrent courses) which open from the ridges are usually dry. Farming, with millets as the chief crop, depends on tanks and wells. There is much open, thorny forest. Westward the area grades into the *Thar*, almost total desert in parts but with some poor grazing. The Aravalli upland and the Thar together comprise the greater part of Rajasthan, the modern administrative unit which corresponds broadly with the former Rajput states whose proud rulers were usually able to preserve a high degree of independence. Economic improvement depends on irrigation. In the north much has already been done by bringing water along the old Ghaggar channel (see page 110). Along the Luni, which flows towards the mud-flats of the Rann of Cutch, underground water might be sufficient for tube-well farming. In the south and at Lake Sambhar, near Jaipur, salt accumulations have assumed national importance since the loss of Salt Range minerals to Pakistan in 1947. Jaipur (population 300,000), Jodhpur (population 181,000) and Bikaner are all former state capitals and trading centres on caravan routes.

Malwa, drained by the Chambal, has a southern portion on the Deccan lavas where millets and cotton thrive, and a northern part of poorer country. Lying between the arid Aravalli-Thar country and the forested uplands which pass eastward into the Chota Nagpur plateau, it commands the best routes from Delhi and Agra to the west-coast ports. In the north is the spectacular fortress of Gwalior, which crowns a rock two miles long with cliffs rising 300 ft. from the plain. There is a plan for a multi-purpose development scheme on the Chambal.

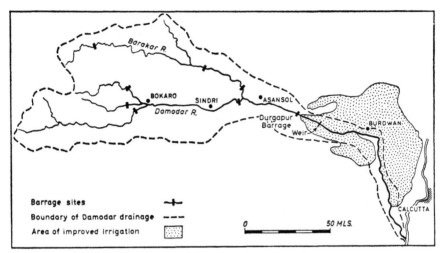

Figure 22. *The Damodar valley scheme*

The plateau of *Chota Nagpur*, about 300 miles from east to west, rises sharply from the Ganges plains to a general level of 1,300 ft. The western part is forested, but most of the eastern half is cultivated. Rice is grown, generally on the floors of minor valleys and depressions where rain water, draining from higher levels, can collect on the paddy fields, though there are instances of rain-fed paddy fields on the crests of interfluves. Over half the farmland carries *tanr* crops, i.e. dry crops such as maize, gram and oilseeds (*tanr* is a local name). *Tanr* land generally lies at higher levels than the paddy, extending from broad valley sides on to the higher surfaces of the plateau. Population densities reach 300 per square mile.

The several mining towns along the fault trough of the Damodar, of which the largest is Asansol, provide employment for surplus rural workers. The Damodar valley hydro-electric scheme (Figure 22) has brought more electricity to the towns, and established a sulphate of ammonia plant at Sindri. To the villages in the remote catchment areas, where several of the dams have been built, the Damodar Valley Corporation has introduced new ideas of farming and forestry which will ensure better crops and less soil erosion. The people of several villages which were inundated by the reservoirs have cleared new land in nearby forests, taking care over surface run-off and the provision of compact holdings. Soils generally are not very good. This is one of the few areas of India where virgin land remains to be brought into cultivation.

Other towns are Ranchi, an educational and administrative centre, and Hazaribagh. Bauxite from the Ranchi area is smelted at Muri. The Damodar valley carries the main rail and road link between Delhi and Calcutta.

Maharashtra has striking geographical individuality as the country of the Marathas (see pages 86, 93). It corresponds broadly with the Deccan lava outcrop south of the Satpura Hills (Figure 21). Lying in the rain shadow of the Western Ghats, this area gets only 20 to 30 in. of rain, and there is hardly any irrigation except for sugar. The *regur* soils yield good cotton, and the main food crop is millet. Rural population densities are about 200 per square mile. Poona (population 598,000), a cultural and administrative centre, was formerly an important British military station. Nagpur (population 644,000) is an industrial town. Sholapur (population 338,000) is one of the great cotton towns of India.

Lighter soils, greater scarcity of surface water and the impressive tors which dominate the landscape contribute to the scenic change on crossing the lava boundary towards the east. With rainfall still below 35 in., millet is the chief crop, followed by maize and oilseeds. There is very little rice. Irrigation, not greatly developed, is mostly from tanks. The dam on the Tungabhadra tributary of the Kistna, one of the major engineering feats of independent India, will ultimately bring water to over a million acres. Hyderabad (population 1,100,000), capital of the former independent state of the same name, is an administrative city with some industry.

The largest expanse of comparatively low ground in the peninsular uplands is the plain of *Chhattisgarh* in the upper Mahanadi Basin. Wetter than areas farther west, and with tank and canal irrigation, it is rice-growing country.

The peninsular uplands present to the west a steep, forested scarp face, the *Western Ghats*, about 3,000 ft. high in the north and increasing to 5,500 ft. in the south. Rainfall is very heavy on the western slopes. Several streams in the neighbourhood of Bombay have been harnessed to give power for the city and electric traction on local railways.

Mysore, and the Nilgiri and Cardamom Hills. In Mysore the plateau rises from around 1,500 ft. in the north to 4,000 ft. in the south. Rainfall increases southward to about 35 in. Agriculture is more diverse than in Maharashtra. Rice, millet, gram and oilseeds are raised with the help of

tank irrigation. The Nilgiri plateau reaches 8,000 ft. and has a high rainfall. It is mostly forested but there are tea and coffee estates. Ootacamund
(7,000 ft.) is a hill station and holiday centre. The Cardamoms are also
forested, and the slopes carry tea, coffee and rubber plantations. These
three areas are grouped under one heading because they all lie partly
within the catchment of the Cauvery river, which has been greatly
exploited for hydro-electricity. There are four major dams: one built as
long ago as 1902, and another, the Mettur, 176 ft. high. While the combined potential of the four is less than half that of Bhakra-Nangal they
have assisted industrialization over a wide area. Electricity is supplied to
the Kolar goldfields of Mysore state, Mysore city, the light industries
including motor and aircraft assembly plants at Bangalore (population
905,000) and to various industries in towns on lower ground to the east
and south, including Trichinopoly, Madura, Salem and Coimbatoire.
The headwaters of the Periyar river which flows to the west coast have
been impounded and drawn off through a watershed tunnel to irrigate
land on one of the Cauvery tributaries.

(3) THE PENINSULAR COASTLANDS

In the north-west the tidal mudflats of the Rann of Cutch pass northward
into the Thar. The peninsula of *Kathiawar*, built mainly of Deccan lava, is
mostly below 600 ft., but peaks in one of the ranges reach 2,000 ft. Rainfall
is low. Wheat, millet and cotton are raised by tank and well irrigation.
Port Okha, flourishing under progressive management, has chemical and
cement works, and there is a remarkably close net of metre-gauge railways.
   East of Kathiawar, the alluvial tract along the Sabarmati river and the
coastland south as far as Damão comprise *Gujarat*. The drier north is
millet and cotton country. Southward, with increasing rainfall, there is
more rice. Ahmadabad (population 1,150,000), one of the great textile
towns of India, owed its early importance to its position at the lowest
easy crossing of the Sabarmati. Broach and Surat (population 288,000),
both active ports in the sixteenth century, declined on account of the
silting of the Narbada and Tapti estuaries and the increasing size of ships.
Both have textile and engineering industries.
   Southward, the strip between the Ghats and the Indian Ocean is about
50 miles wide. In the north only a small proportion of the area is alluvial.

Much is jungly, lateritic terrace at about 200–400 ft. Rainfall increases to 100 in. and more. Rice is grown in valley bottoms, and since Partition jute also. Bombay on its island site is a city of about three million people with textile, engineering, chemical and oil-refining industries. South of Bombay, the proportion of alluvial ground to less useful country increases, and the coconut palm becomes as prominent as the paddy fields. *Kerala*, the coastland from Calicut southward, is described by Spate as 'physically and culturally one of the most distinctive regions of India'. There is a lagoon coast, and intensive paddy cultivation, with rural population densities up to 4,000 to the square mile. Cochin, on the broad-gauge railway, is a growing modern port. Hydro-electricity from hills inland is used at Cochin and by the aluminium smelters of Alwaye, and for pumping water from coastal paddy fields. There are many minor ports with roadstead anchorages for small coastal shipping.

Of the three former Portuguese areas on the west coast, the largest is Goa. There is a modern port, with remarkable ruins of churches dating from the sixteenth century. Many Goans emigrate from their crowded homeland to make their living in professional and clerical work, especially in Bombay.

The east-coast lowlands are wider than those of the west, and there are the deltas of the Mahanadi, Godavari, Kistna and Cauvery, all of which have easily irrigated alluvium supporting dense populations of rice farmers. Rainfall decreases southward. In the flat country around Madura, where rainfall is only about 30 in. a year, every practicable site has been bunded to retain water. Of the coastal towns, Vizagapatam is the main shipbuilding town of India. This is related to the fact that it is the nearest *ocean* port to the steel works of Jamshedpur. Of Madras (population 1,730,000), with its poor artificial harbour, Spate writes:

> Strictly speaking its site has no advantages whatever, except that its position midway between Pennar and Ponnaiyar probably counts for something. But in the general setting of early English activity along this coast some commercial and administrative centre was bound to develop, and any site once developed was bound to maintain its position since, if it had no special values itself, no possible rival had any either, and a going concern had a certain pre-emption. But it is obviously solely a matter of history that Madras is today more important than Pondicherry.

As recently as the early eighteenth century a substantial proportion of the Ganges water reached the sea through the western distributaries of the delta. Today the main outlet of the Ganges and Brahmaputra is through the eastern distributaries. The western channels, except the Hooghly, are silted. The western part of the delta, immediately east of the Hooghly, is badly drained, malarial country where old river courses are marsh for most of the year. There is no annual inundation by silt-laden flood water. In contrast, in the eastern part of the delta in East Pakistan there are wide channels, the annual flood brings fertile silt to much of the farmland and better drainage makes for healthier conditions.

West Bengal is a political unit: the part of the former province of Bengal awarded to India at the Partition of 1947, with some later minor adjustments of borders in the north. There are four main divisions:

1—east of the Hooghly, a long strip about 50 miles wide, inefficiently drained by decaying channels. At the seaward edge are the swamp forests of the Sundarban;
2—west of the Hooghly, a wider lowland, much of which is a lateritic shelf with scrub forest bordering the Chota Nagpur plateau;
3—to the north, a portion of the Ganges-Brahmaputra *doab*, some of which is lateritic older alluvium;
4—on the northern border, the forested Himalayan foothills.

In the first three, rural densities of population are high except on lateritic ground, though agricultural opportunities are not as good as in the eastern part of the delta. Rice is the dominant crop, but there is nothing like enough to support the rural areas and the urban population of Hooghlyside. Nor is there much scope for extending agriculture, except to a small degree on the Ganges-Brahmaputra *doab*. Such an extension might not result in more food. Over three quarters of the rice is grown in broad valley floors as an *aman* crop, i.e. it is planted in June and harvested between November and January. Quick-ripening paddy with lower yields is grown on slightly higher ground as an *aus* crop, i.e. it is planted in the pre-monsoon rains in April and harvested between July and September. Dr Chakraborty of Calcutta has shown that the extension of *aus* on valley sides on the western border of the area has had an adverse effect on the hydrology of the best *aman* ricelands of adjacent valley bottoms, with resulting poorer yields.

The sharp southward bend of the Damodar dates from the flood of 1770 when the river changed from its eastward course to enter the Hooghly south of Calcutta, over 80 miles below the old confluence. In this low ground west of the Hooghly variability of rainfall is enough for irrigation to be an asset; the more secure water supply from the Damodar valley scheme is a big help to farming.

The hills of the Himalayan border, mostly forested, have been cleared in places for tea estates. Some valley sides have been terraced for dry grain crops and irrigated rice. Darjeeling (7,376 ft.) is a holiday centre. In normal times Kalimpong, at the Indian end of the main trade route from Lhasa, is a market for Tibetan wool.

Hooghlyside. Calcutta (population unknown but certainly over three million), the British capital in India from 1773 until 1912, was founded about 1692 by Job Charnock of the East India Company, 100 miles from the open sea, on the east bank of the treacherous Hooghly channel, which has a 10 ft. tidal bore running for several miles north of the city. The port has modern docks and handles about a third of the oversea trade of India. The *site* has no natural advantages, though the *location* on the western side of the delta was chosen by the British as the best available gateway to the Ganges plains. There are many public buildings of architectural merit. The city centre, with the wide-open space of the Maidan, is well planned but outer areas are among the most squalid in Asia. Calcutta is a commercial, administrative and university city. The creation of East Pakistan drew off some of its trade to Chittagong, but that loss is more than outweighed by prospects of economic development in the remainder of the hinterland. The industries, mainly jute, paper and engineering, are in the riverside towns which form an almost continuous urban belt for 20 miles northward and 10 miles southward along both banks of the river. Howrah, on the west bank opposite Calcutta, is an industrial town and terminus of the railways from Delhi and the peninsula.

(5) ASSAM AND THE EASTERN BORDERS

The atlas maps show four physical constituents:

    1—the Brahmaputra valley, 400 miles long and 60 miles wide;

    2—the mountainous border on the Himalayan edge;

3—the Shillong plateau;

4—the hill country of the Burma border.

The Brahmaputra valley is thinly populated by rice farmers; on the bordering hill slopes there are tea estates. Dhubri, Gauhati and Tezpur are small towns and river ports sited where old hard rock conveniently projects from beneath the alluvium. The Shillong plateau, lying between 4,000–6,000 ft. and mostly forested, has a small population of shifting cultivators. Shillong, the capital of Assam, has only 53,000 people. The Himalayan border is wild and little-known country, while the ranges of the Burma border are populated by shifting cultivators such as the Naga tribes. The Brahmaputra is navigable. There are metre-gauge railways to Chittagong and West Bengal. At Digboi in the north-east there are small but productive oilfields. A 720-mile pipeline leads to a refinery at Gauhati, and on to a large new refinery at Barauni on the Ganges near Patna.

Looking at Assam and the eastern borders as a whole, the hill and plateau portions have little to offer. The Brahmaputra valley could support many more people if drainage and river control were undertaken; there is some scope too for hydro-electric schemes. But all this would require capital expenditure in an area remote from the centres of political power. During World War II many airfields were made, railways were improved and several good roads were built in the hills on the Burma border. Two considerations dominate the question of whether the Indian government should undertake vast development schemes. The first is economic; would it be worth while or would the money be better spent elsewhere in India? The second is political and strategic. A large and relatively empty area of good land in overcrowded Monsoon Asia is a temptation to neighbouring powers. As early as 1900 the British government felt this area might be threatened by Chinese overspill through Tibet. More recently, in the arguments preceding the Partition of 1947 the Pakistan faction claimed Assam for their excess population of East Pakistan.

Manipur on the Burma border and Tripura on the border with East Pakistan are two former states now governed directly from Delhi on account of their strategic situations.

(6) THE HIMALAYAS

In this area of tremendous relief, so dissected by gorges that there is hardly any level ground, the topographical and human details are immensely complicated. The broad characteristics are as follows: west of the Kosi river the Siwalik Hills, formed in unconsolidated sands and gravels, are a distinct feature, rising to 5,000 ft. and separated from the main Himalayan front by sandy lowlands (*duns*) up to 20 miles wide. East of the Kosi there are no Siwaliks. The harder rocks of the main ranges rise abruptly from the plain. For the most part rainfall is high, but falls below 30 in. in the local rain shadows of the west, where gorges such as those along the Sutlej are shielded by higher ranges from the summer monsoon. The mountains are wooded, with rain forest and monsoon deciduous forest on lower slopes, succeeded by evergreen oak and other species around 4,000 ft. Timber-felling is locally an important industry. Coniferous trees begin around 8,000 ft. and merge into rhododendron forest which reaches 13,000 ft. Above that level slopes are rocky, with patches of alpine grass, some bush rhododendron and stunted juniper. In the west, where winter snow occurs as low as 5,000 ft., the lower limit of permanent snow is between 16,000 ft.–17,000 ft. In the east it is generally lower, about 15,000 ft., though exposed places above that level are kept clear by strong winds. The glaciers reach down to 13,000 ft.

Approaching from the plains, one enters first the *terai*, a strip of marshy, malarial jungle, hilly in parts and perhaps 20 miles wide in the east but much narrower in the west. For the most part empty of people but full of game, it offers some prospect for future agricultural settlement but at high cost in clearing, drainage and river control. Northward, the mountain slopes are densely forested to 3,000 ft. Above that level many valley sides have been cleared and terraced for agriculture. Water is brought to terraced fields by contour leats and carried across ravines by crazy but effective guttering made of split bamboo. In areas of higher rainfall in the east the annual damage to terraces by storm and landslip requires continuous repair work. Villages are few. Most of the people live in isolated farmsteads. Paddy is grown to about 5,000 ft., maize does well up to 7,000 ft. Dry rice and wheat are raised in unterraced clearings. Barley is grown on small alluvial cones up to 12,000 ft., though at that altitude the season is too short for the grain to ripen and the crop is harvested as hay. At these higher levels stock-raising is more important than agriculture, and people are fewer. Goats and sheep are grazed on high summer pastures and driven in winter to lower levels where sedentary farmers are prepared to accept them on their land to get the manure. (By tradition these transhumant people of the higher slopes were also traders who carried bazaar goods from India across the high passes to Tibet.) Lead, copper and zinc are widely scattered in small deposits, and systematic surveys might well reveal greater mineral wealth; but the intermittent working of copper ores in Sikkim has been greatly hindered by transport difficulties.

During the British period this Himalayan border was peaceful. Had it not been so, the maintenance on this 1,200-mile border of garrisons, similar to those on the mere 500 miles of the north-west frontier, would have been a costly burden. As it was, the Chinese claimed Tibet but they interfered very little in the actual country. Practically all the Tibetan export trade (mostly wool) moved southward on pack-mules to India. The little that went eastward into China risked attack by bandits whom the Chinese were unable to control. From Lhasa, Calcutta was the most accessible port. Over half the trade came by one track, the Lhasa-Kalimpong route, which crossed from Tibet into the Indian state of Sikkim and thence to Kalimpong in the hills of northern Bengal.

Europeans established the fashion of going to the hills in the hot weather. Among the hill resorts are Simla, Musoorie, Nani Tal and Darjeeling, all between 5,000 ft.–7,500 ft. Until recently the only roads were those linking these towns with the plains. There are tea estates in the east around Darjeeling and in the foothills in Assam.

BORDER PROBLEMS

When China was a weak power, with slender military resources thinly spread, the Himalayas with the Tibetan plateau beyond seemed a safe natural defence for India. The emergence of the Chinese People's Republic as a strong power with a population of over 600 million has forced on the Delhi government a revaluation of the security of this northern frontier. The Chinese invaded Tibet in 1950. Motor roads were built from Chinese railheads to Lhasa and to several points on the Indian border. Tibetan export trade was diverted to Chinese markets. These events, followed by Chinese claims to mountain country long regarded (at any rate by the western world) as Indian, and the flight to India in 1959 of the Dalai Lama chased by 50,000 troops, emphasize the threat of Chinese infiltration of Indian territory foreseen by the British 60 years ago (Figure 23). The Indian government is building roads from the foothills to strategic areas on the border, but the difficulties are great. Once constructed, a road on the drier Tibetan side of the Himalayas is easy to keep in repair, but the Indian roads must run through the rain-drenched gorges of the southern side, where the rock is deeply weathered and land-slips are a constant menace. The Indian position is complicated by the presence on the border of the two independent states, Nepal and Bhutan, in which Russia, China and western countries compete with offers of economic help.

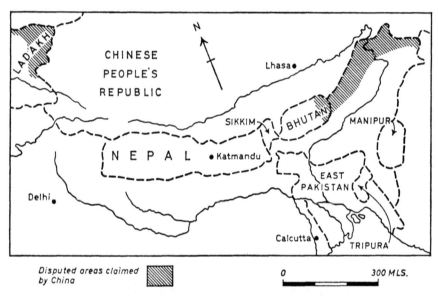

Figure 23. *Disputed territory on the Himalayan border*

Nepal. There are about eight million Nepalese, mostly Buddhist peasants farming on forested mountain slopes and also in valley bottoms where these are not malarial. The more densely settled parts are separated from the Indian plains by forested hills with poor sandy soils and a belt of *terai* jungles notorious for malaria. This physical seclusion helped the ruling Rajput family to maintain their independence. Until recently the capital, Katmandu, was accessible from India only by a difficult jungle track. There were treaty arrangements by which Europeans could visit Katmandu only by permission of both British and Nepalese authorities, and that permission was rarely granted.

Although there is plenty of unoccupied mountain land which could be terraced for cultivation, many Nepalese have settled in the *terai*. Others have taken up land in southern Bhutan, where they form a powerful minority. Many again find work in India. The Gurkha Rifles, consisting almost entirely of Nepalese men, is part of the British army. The British government pensions to retired Gurkhas (about £700,000 in 1960) are an important source of foreign exchange. With their annual domestic revenue of only about four million pounds, the Nepalese could do little to develop their country, but foreign aid over the last 10 years to the extent of 50 million pounds is showing results in new roads and small hydro-electric schemes. The plans for a great dam in the Chatra gorge on the Kosi river would bring power and irrigation to the Nepalese *terai* as well as to India.

Bhutan (population about 700,000) is more backward than Nepal. The small surplus of rice once found a good market in Tibet, but this trade ceased with the Chinese occupation in 1950. The Indian government, realizing the strategic weakness of a roadless state occupying 200 miles of the frontier, has encouraged the Bhutanese to look southward for economic opportunity. Help is given in education, especially in the teaching of Hindi, which enables Bhutanese to take employment in India. New roads, financed and planned from India, are bringing the central and southern part into easy communication with West Bengal.

Sikkim. Wedged between Nepal and Bhutan is the state of Sikkim, a roughly rectangular area about 70 miles from north to south and 40 miles wide, which contains the eastern slopes of Kanchenjunga and the headwaters of the Tista. Here, through some of the world's most spectacular relief, runs the Kalimpong-Lhasa trade route, the easiest track from India to Lhasa. The route, which was improved and graded by Colonel Younghusband's military mission to Lhasa in 1904, is more practicable than any other for the construction of a modern motor road. In 1950, at the request of the ruling maharajah, the Indian government undertook responsibility for defence and communications.

II

Pakistan and Kashmir

In the Indus lowland of West Pakistan, agriculture, based on modern perennial irrigation, achieves a higher material living standard than in most other parts of the sub-continent. There is considerable scope for the extension of irrigation to potentially productive alluvial land. In good years there is a local food surplus, but this is outweighed by shortage in East Pakistan, where 55 per cent of the total population of Pakistan is concentrated in conditions of extreme poverty on a mere 15 per cent of the total area of the state. Overcrowding and malnutrition in East Pakistan are urgent social problems; in the government plans for the economic development of Pakistan as a whole further irrigation and factories for processing of agricultural products are the main items. No great industrial development is contemplated, as mineral and power resources are slight, apart from the salt deposits of the Salt Range and the natural gas of Sui in Baluchistan. Since 1947 economic progress has been hindered by the diversion of a high proportion of the slender revenue to the maintenance of armed forces on the borders with India and in Kashmir. (see page 147)

Today 85 per cent of the population of Pakistan is Muslim, but in 1947 the chief towns of the Indus area had slightly more non-Muslims than Muslims. Among these non-Muslims (chiefly Hindus and Parsees) were many rich and prominent business men who might have played a useful part had the social climate encouraged them to stay. Many of them sought better prospects in India, while the Muslims who moved in from India were mostly destitute, adding another burden to the already harassed government.

The division of the country into two zones separated by 1,000 miles of Indian territory is a permanent economic and political handicap. The sea route from Karachi to Chittagong is 2,700 miles. Use of the direct land route has been hindered by high transport costs on railways already overloaded and by political tension between Pakistan and Delhi.

The atlas maps show two broad physical divisions—the Indus plains and the higher ground of the west and north. The Indus plains fall into two parts: the lower Indus valley, in which the Indus receives no major tributaries; and the Punjab plains, the northern portion, drained by the Indus and its four tributaries (Jhelum, Chenab, Ravi and Sutlej). The fifth major tributary of the Indus drainage, the Beas, is entirely within India. The higher ground of the west and north can be subdivided as follows: Baluchistan, the north-western hills, the Potwar plateau and the Salt Range. Baluchistan is also a political division, and as such its boundaries include areas of plain bordering the Indus lowland.

The lower Indus valley is a plain about 100 miles across and 400 miles long, bordered on the east by the Thar Desert and on the west by the Kirthar and other ranges. Most of the ground to the east of the present course of the Indus is *khadar*. To the west, most of the land is *bhangar*. There are two outcrops of limestone in the plain, one rising to 400 ft. at Sukkur, the other reaching 250 ft. and providing a site for the city of Hyderabad (Sind). While the river course has shifted a good deal over the *khadar* areas even during recent centuries, the more permanent channel through the limestone at Sukkur provides a good site for the Sukkur (Lloyd) Barrage, which was completed in 1932. That even the barrage site is not completely secure is indicated by an abandoned channel in the limestone only four miles away from the present river course. In an exceptional flood the river could plough a new bed and leave the barrage high and dry. To minimize this risk, earthworks upstream are carefully maintained. The West Nara (Figure 24), a distributary which takes off from the Indus below Sukkur, might once have been the main river channel. In early times the East Nara was probably the lower course of the Sarasvati (see page 108). The delta is a mangrove-fringed waste, merging southward into the mudflats of the Rann of Cutch.

The lower Indus valley is remarkable for its fierce heat. Shade temperatures at noon exceed 100°F. for most of the summer. Frosts occur in December and January. Figures of mean annual rainfall (Karachi, 8 in.)

mask the fact that many years are rainless, while infrequent local downpours might amount to double the annual average in a few hours. For
example, on 6th September 1959, at Karachi, 5 in. fell in 18 hours.

Share-cropping on large estates is the usual farming practice. All the
cultivated land is irrigated. Acreages of the two main crops, wheat and
rice, are about equal, with cotton not far behind. Millets, oilseeds and
dates are also grown. Some 36,000 miles of main channels and distributaries carry water from the Sukkur Barrage to irrigate 5·5 million acres,
only 0·75 million acres short of the entire irrigated area of Egypt. A

Figure 24. *The lower Indus valley and part of Baluchistan*

further 2·8 million acres, mostly under rice, is watered from the Kotri Barrage completed in 1955.

North of the delta, Karachi (population 1,913,000) grew as the port for the Indus plains. With the extension of irrigation in Sind it acquired an export trade in wheat and rice. Other commercial functions came with the status of national capital in 1947. The airport has long been a staging post on world air routes. Despite the growing factory estate along the road to the airport, and the natural gas brought by pipeline from Sui, industrial development is slight.

The Punjab plains. The *khadar* areas along each of the main rivers are separated by the higher *bhangar* of the *doabs*. The alluvial cover is generally very thick, but an indication of a northward continuation of the hard peninsular rocks below the alluvium is seen in the quartzites, slates and rhyolites of the isolated Kirana Hills which rise 1,000 ft. above surrounding country south of the Chenab.

The northern fringe of the Punjab plains next to the Himalayan foothills has a high water table and about 35 in. of rain, including a useful contribution from winter depressions. Sugar cane, *rabi* wheat and *kharif* millet, maize and cotton are raised. Well irrigation is prominent, land holdings are very small and rural population densities are high.

Southward, rainfall decreases rapidly and agriculture becomes increasingly dependent on irrigation. Lahore gets about 20 in., while the southern half of the area receives on average barely 10 in., and that notorious for its variability. The *khadar* areas, watered from wells and inundation canals, have a long history of cultivation, but the higher *bhangar* of the *doabs* was neglected until long perennial canals were built. The first of these canals dates from the fourteenth century. But little was done until the British undertook large schemes after 1857. The headworks of the Upper Bari Doab canal were completed on the Ravi in 1859. Most spectacular of all was the Triple Canal Project, completed in 1917 (Figure 25). In the first place the Upper Jhelum canal was built to irrigate some 350,000 acres before discharging into the Chenab. This discharge contributed to irrigation canals downstream on the Chenab, so that water upstream of the discharge was available for use in the second part of the project, the Upper Chenab canal, which irrigates 650,000 acres on its eastward course towards the Ravi. On reaching the Ravi, the surplus water is carried across that river on the top of a barrage and directed into

the third part of the scheme: the Lower Bari Doab canal, which waters
the near-rainless country between the Ravi and Sutlej. This elaborate
arrangement for taking Jhelum water 50 miles eastward to the Chenab,
and Chenab water 200 miles southward over the Ravi, reflects the
physical geography in two ways. First, as noted on page 106, the right
banks are generally higher than the left banks, so it is easier to provide
barrages to divert water into left-bank canals. Second, the Ravi has the
smallest catchment and volume of all the Punjab rivers, and its water
was already almost fully utilized by the Upper Bari Doab canal. New
irrigation downstream and south of the Ravi had to call on supplies from
other rivers. The Sutlej water, which was near at hand, was mostly needed
for new schemes south of that river.

Figure 25 also shows the orientation of the main canals throughout
the Punjab plains. Apart from the three main waterways of the Triple
Canal Project, they generally run from north-east to south-west. In the
north-east they water land which was already thinly settled, so the new
irrigation was fitted into the existing rural landscape with all its agrarian
problems. In the south-west, and especially in the districts of Lyallpur and
Montgomery, they brought water to arid, uninhabited country; there
were no rural settlements or rights to land to hinder the planners, so the
new villages were laid out on a rectangular pattern and farm units were
large. Settlers from crowded areas to the north and east were selected
carefully. These canal colonies are today the most thriving rural areas of
the sub-continent. In addition to the usual crops of wheat, cotton, maize,
rice, sugar cane and oilseeds, there is a substantial acreage of fodder crops
which support cattle in relatively good condition. Diet includes milk and
eggs, and the nutritional standards are higher than in any other rural area
of Pakistan or India. One drawback spoils the picture of prosperity:
seepage through the beds of the canals means that much water is lost. The
water table close to the canals has risen in places so much that water-
logging of the topsoil prevents cultivation. Continuous pumping, an expensive cure, is sometimes effective, but large areas are beyond this treat-
ment. The only permanent remedy, lining the canals with concrete, is
virtually impossible because lengths of canal cannot be withdrawn from
service for the necessary time.

The dispute of 13 years' standing between Pakistan and India over the
division of the waters of the Indus drainage was resolved in 1960. India
was granted the flow of the three eastern rivers Sutlej, Beas and Ravi,

while the waters of the Indus, Jhelum and Chenab are reserved for Pakistan. This arrangement involves long new canals in Pakistan to bring water to areas near the Beas and Sutlej previously watered by canals from

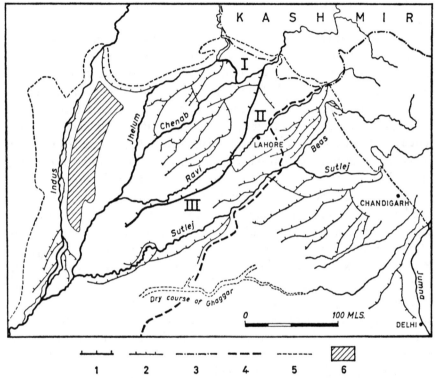

Figure 25. *Punjab irrigation: the Triple Canal Project* Explanation of key: 1—main canals of Triple Canal Project (I—Upper Jhelum Canal, II—Upper Chenab Canal, III—Lower Bari Doab Canal); 2—other canals; 3—boundary of Kashmir; 4—Indo-Pakistan boundary; 5—approximate junction of hills and alluvium; 6—the area of the Thal and Sind Sagar irrigation scheme begun in 1953

those two rivers. The change is to be achieved gradually over a period of years with the help of loans from the World Bank. Ultimately India will be able to draw heavily on the three eastern rivers for irrigation in the Indo-Gangetic divide (see page 110 and Figure 19).

In 1947 the only large areas of the Punjab plains in Pakistan that remained unirrigated were the desert and scrub between the Indus and

Jhelum (the Thal and the Sind Sagar *doab*), together with a strip of alluvial ground west of the Indus. Water is being brought to these areas by canals from the Taunsa Barrage, 80 miles below 'Kalabagh on the Indus, and by the Thal Project, which draws Indus water from head-works at Kalabagh. These two schemes will ultimately irrigate 3·5 million acres.

The agricultural products and natural resources of the plains are not such as to encourage industry. Most of the cotton, after ginning in local mills, is exported. There is no local fuel. Small amounts of lignitic coal are brought from the Salt Range mines. Power lines from the Mandi hydro-electric station on a tributary of the Beas in India reach several Pakistan towns, and small amounts of electricity are generated at the irrigation headworks within the Punjab plains. A pipeline from Sui in Baluchistan supplies natural gas to Multan and other towns. Processing of food and other crops, light engineering and manufacture of various consumption goods occupy small factories in many towns. The rugs, blankets and carpets of Punjab craftsmen working in small groups have a world reputation for high quality. Lahore (population 1,300,000), centre of Muslim learning, was the capital of the former province of Punjab in British India. Most of the older towns grew as strong points and trading centres: Jhelum and Gujranwala on the route from the north-west, and Multan on the approaches along the Indus from the south-west. Lyallpur and Montgomery are modern towns that owe their growth to the irri-gation schemes of the present century. The Punjab plains have a good railway network. The lines converge southward on Lhodran, whence one line continues to Sukkur and Karachi.

Baluchistan. Some of the Sulaiman ranges on the eastern border exceed 10,000 ft. and several are high enough to induce a rainfall sufficient for forest growth. But, for the rest, rainfall is low and unreliable, tempera-tures are extreme and the vegetation (where any exists) is usually coarse desert grass with scrub along the stream beds. The interior plateau, which lies between 2,000-3,000 ft., is practically rainless sand desert. With the help of *karez* irrigation, wheat, millet, apricots, apples and grapes are grown on alluvial stretches in the eastern hills. Date palms provide an additional staple food in valleys along the Makran coast. Most of the farmers also keep sheep, and there is a good deal of seasonal movement of livestock into the edges of the Indus lowland to avoid the harsh winters.

Camels are much used for transport. The population density overall is about five persons to the square mile and large tracts are empty.

The Makran coast provides a way into the Indus lowland from the west. The coast route itself is waterless over great distances; but some of the longitudinal valleys that run parallel with it, and up to about 100 miles inland, offer better going in parts through farming and grazing country fed by ground water. Alexander the Great and part of his army withdrew from India along these Makran routes in the August of 325 B.C., and many of his men died from heat, thirst and exhaustion. In the eighth century A.D. the first Muslim invaders of India used the same tracks. Remains of gardens, towns and irrigation works suggest more humid conditions prevailed at that time.

Another entry to the Indus lowland through Baluchistan is by the Bolan Pass. Spate points out that although the Makran and Bolan routes are relatively easy, they never achieved the importance of the routes in the Khyber area, possibly because further progress into India was blocked from their direction by the Thar Desert, while the Khyber routes open directly on to the Punjab passageway to the Ganges plains.

There was no great economic justification for the railways from the Indus plains by the Bolan and Harnai passes to Quetta, the extension to Chaman on the Afghan border, and the branches north-east to Fort Sandeman and westward to the border of Iran; their significance was mainly strategic. Reaching to within 100 miles of Kandahar, the Chaman line carries some Afghan trade, including exports of carpets and wool and imports of consumption goods. There is a small production of Tertiary steam coal at outcrop workings near Quetta and from adits at the railway colliery in the Harnai Pass; but the seams are thin and difficult to work, and the planned increases in output will make little impression on the fuel situation in Pakistan. More significant for the future are the chromite workings in the Zob valley reached by road from Quetta, and sulphur in the Chagai Hills on the Afghan boundary. The natural gas field of Sui, which supplies Karachi and Multan, is on lower ground, where the Sulaiman structural alignments curve westward. Quetta, the only town of any account, grew as a military and administrative base in the nineteenth century, but acquired only minor commercial importance.

The north-western hills (Figure 26) are described by Spate as the ragged fringe of Afghanistan. In the north is the great wall of the Hindu Kush.

Southward the land is a tangle of ranges and narrow valleys. Rainfall is higher than in Baluchistan, more of the ranges are wooded and agricultural opportunities are slightly better, but aridity is still the dominant characteristic. The routes through the Khyber area, which converge on the Vale of Peshawar, were the principal means of entry for the successive invasions of India by people from the north-west. Spate warns that the Khyber Pass itself has attracted more than its fair share of attention. Easier routes from the Kabul river north-eastward along the Kunar and over cols into the Swat valley and thence downstream towards Peshawar were certainly used by Alexander the Great and Babur (see pages 57, 61).

The north-western hills are the country of the Pathan tribes, all Muslims and of Pushtu speech. They keep sheep and goats, and cultivate fields watered by *karez* and minor irrigation channels. This poor living was traditionally supplemented by raiding the richer lowlands to the south and plundering traders on routes to Afghanistan. Yet the tribes failed to combine into larger and more powerful political units, a fact probably related to the roughness of the country and the absence of a good agricultural base from which the whole could be controlled. Their capacity for anarchy and unrest was so great that some protection for the Indus plains was thought essential by the British in the nineteenth century. Kabul and Kandahar were occupied by British forces during the Afghan Wars (1839-42 and 1878-80), but to pacify and administer all the hill country between the Indus and those two cities would have involved intolerable expense on road construction and garrisons in wild and hostile country which could yield no tax revenue in return; the problem was further complicated by the prospect of Russian penetration through Afghanistan, and the absence of an agreed border between Afghanistan and India. The present border with Afghanistan, *the Durand Line*, was negotiated by Sir Mortimer Durand in 1893. Of the north-western hill country thus included in India, only areas adjacent to the Indus lowland were administered. In the remainder, known as 'tribal territory', a partial and uneasy authority was maintained by a mixed policy of financial subsidy and military threat, which at least had the merit of relative cheapness. British garrisons held key points, while local men were recruited and paid well for military duty and construction work on roads and railways. This proved an important source of income to the tribes, so they saw to it that there was enough unrest to justify the British staying on (see O. H. K. Spate, *India and Pakistan*, pages 434-50). In 1947 this intractable problem

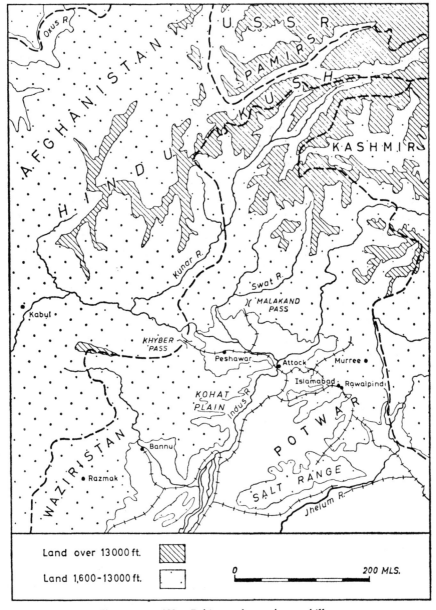

Figure 26. *West Pakistan: the north-west hills area*

was inherited by Pakistan. The financial subsidies were maintained but the military posts were abandoned. Any lasting solution must involve reduction of the population dependent on the agricultural resources of the hills, and the provision of employment in more favoured adjacent areas.

In the north-western hills are three prominent lowlands, all draining to the Indus: the Bannu plain, the Kohat plain and the Vale of Peshawar. The first two are of little account at present but potentially valuable if more irrigation can be provided. The third, the Vale of Peshawar—some 90 miles across and drained by the Swat and Kabul rivers—has a flourishing irrigated agriculture, with crops of wheat, maize, sugar cane, rice, tobacco and cotton; some of this irrigation water is brought from the upper reaches of the Swat through a tunnel under the Malakand Pass. The city of Peshawar (population 152,000) dominates the main trade route to Afghanistan. The Malakand hydro-electric power station supplies current for a growing range of light industry, and the multi-purpose scheme involving a dam at Warsak on the Kabul river is the basis of future economic progress.

On the *Potwar plateau*, rising to 2,000 ft., soils are good but the rivers are incised so that irrigation is difficult. Agriculture is almost entirely dependent on rainfall, which reaches 25 in. only along the northern border. The countryside is poor and overcrowded. Rawalpindi (population 340,000) dominates the main route into Kashmir. There is a small oilfield at Attock. A new national capital, Islamabad, is under construction adjoining Rawalpindi.

The *Salt Range* has a prominent south-facing scarp dissected into deep ravines. The massive beds of rock salt, worked in open quarries, are the outstanding mineral resource of Pakistan.

Pakhtunistan. The Durand Line, grudgingly accepted in 1893 by the Amir of Afghanistan, cuts across the tribal areas. About 3·5 million Pathans are in Pakistan, and a larger number, perhaps as many as seven million, are in southern Afghanistan. To the Pathan tribes, more concerned with inter-tribal feuds than political cohesion, this was of little significance, but in recent years Afghan influences have encouraged the idea of *Pakhtunistan*, an autonomous state to include all the Pathans. This state would embrace all the Pushtu-speaking parts of the north-western hills area, including Peshawar, Kohat and Bannu. The Indus would become the frontier of Pakistan. If Pakhtunistan came into existence landlocked Afghanistan would be a powerful influence in its internal and external affairs, which explains why the idea is sometimes extended to cover all Baluchistan with the prospect of a port site on the Indian Ocean.

EAST PAKISTAN

East Pakistan is about 300 miles from north to south and 200 miles from east to west. Eighty per cent of that area is alluvial lowland below 50 ft. The only high ground is in the south-east, where the parallel ranges of the Alpine folds in the Chittagong area reach 4,000 ft. In the north-east around Sylhet a few hills rise to 300 ft.

Three great rivers converge on the delta: the Padma, the Jamuna and the Meghna (Figure 27); Padma is the local name for the Ganges, Jamuna is the lower reaches of the Brahmaputra, and the Meghna receives water from the Surma and other tributaries draining from the Assam plateau and the eastern border ranges of India. Even in the dry season the Jamuna is 10 miles wide in places. During the annual monsoon floods new river beds are formed and old abandoned alignments are reoccupied, so that the main channels of the great braided streams change from year to year. Land on one bank might be washed away, while on the opposite bank new land is built as sand and silt accumulate. These new lands, known as *chars*, often have rich soils of new alluvium and are the subject of endless litigation among farmers eager to cultivate them.

Figure 15 conveys an impression of the maze of rivers and distributaries. Figure 27, based on a map by Professor Ahmad of the University of Dacca, shows the main rivers as they are today. According to Ahmad, 'the great intermingling of the mighty rivers Padma and Meghna does not appear to have taken place more than 125 years ago'. Around 1770 the Padma and Meghna reached the sea along separate courses; at that time the present Arial Khan was the main Padma channel. In the floods of 1787 the Brahmaputra abandoned its old course (Old Brahmaputra on Figure 27) but some Brahmaputra water still finds its way along the old bed. The Sundarban, a belt of swamp forest up to 80 miles deep, forms the seaward edge of the delta between the Hooghly and the Meghna. Tidal bores, up to 20 ft. high at times, reach 100 miles inland.

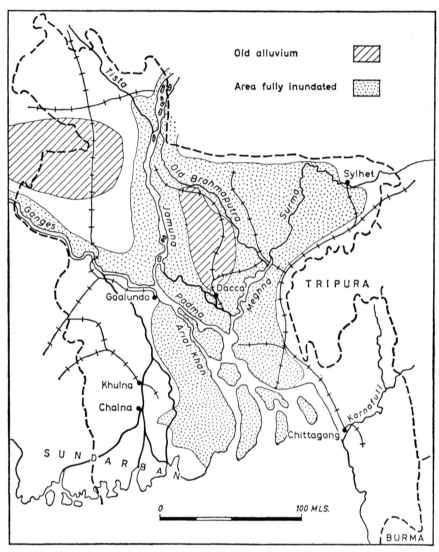

Figure 27. *East Pakistan* The stipple indicates the area in which the fields are
fully inundated by the annual flood. Only the main water courses are shown.
For a more accurate impression of the complexity of the drainage, see
Figure 15, page 99

Agriculture and rural settlement. Mean annual rainfall increases from 55 in. in the north-west to 100 in. along the seaward edge of the delta and 200 in. in the hills north of Sylhet. Much of the lowland receives a layer of fertilizing silt each year from the flood waters of the three rivers. Thus the agricultural potentialities are in general impressive. Some parts are more favoured than others, as indicated by Figure 27. Most productive are the farmlands which are completely inundated every year. They form a belt of country 100 miles across in the south and widening to 200 miles in the north. In this zone the flood covers all except the few railway and road embankments and the sites of farmsteads and the infrequent towns. There are no villages. The farm buildings are in small groups on low platforms of earth just above water level. They are entirely isolated from one another and from the outside world until the floods subside. Two or three feet is the general depth of water, but it might be only a few inches or as much as 10 ft. Throughout this area the general practice is to cultivate all the land every year, and a few fields are double-cropped. Long fallows are unnecessary on account of the silt.

On the borders of the fully inundated area is a less extensive zone which is only partially flooded. Lower levels receive a layer of silt each year. The upper levels, which are just high enough to escape the flood, require long fallow periods.

Two large areas of old alluvium (*bhangar*), the Barind and the Madhapur Tract, stand at a general height of 50 ft. to 100 ft. These are above flood level. Some of the soils show signs of laterization.

On the western border south of the Padma is a portion of the moribund delta. This tract, which continues into India, is the area of silted and abandoned waterways where no extensive floods occur. The lack of annual silt deposits is reflected in the practice of a two-year fallow after each crop. In the hill country of the south-east there are alluvial soils along the lower reaches of the valleys and on the narrow coastal plain. The hills are steep and forested.

90 per cent of the crop land is sown with rice and of the total rice acreage 65 per cent is an *aman* crop. The remainder is almost all an *aus* crop. There is a minute proportion of *boro* rice, i.e. rice sown in low and marshy areas in November and harvested between February and May *Aman* rice thrives in deep water. The *aus* varieties grow to only three feet and are restricted to the higher fields where the flood is unlikely to exceed two feet. Jute, the main cash crop, also does best in shallow water, and is

therefore in competition with *aus* for the higher fields of the inundated area. Small amounts of jute are grown on fields that are never flooded, but the crop is so exhausting that these fields have to be heavily manured. Oilseeds occupy about the same acreage as jute, and there are smaller acreages of sugar cane, tobacco and cotton. Tea is a plantation crop on the hills around Chittagong and near Sylhet.

In the sixteenth century the rich farmlands of what is now East Pakistan produced surplus rice for export to other parts of India. Today the population is so large that in most years the local food supplies are supplemented by imports to maintain even a poor nutritional standard. Rural densities exceed 2,000 per square mile in places, and reach 800 for the plain as a whole. Even on the old alluvium, densities of 500 are common. Only in the Chittagong Hills are the figures below 50. Five acres is thought to be the minimum size for an economic holding in the plain, yet half the farms are less than two acres. Climatic hazards, which would be no more than a nuisance if the population were smaller, bring extreme hardship. In the words of Professor Ahmad:

> If the [early] rains are too heavy, the rice seed is liable to be washed away; if the amount is too little, the younger shoots wither away. If the rains come too late, the seeds cannot be sown in time to allow the young plants to grow high enough to overtop the floods when they come. After the early phase of the Monsoon, once the rivers have overspread the land, it is they who dominate the situation much more than the local rainfall. If the water is too deep, the rice plants are liable to be drowned outright or swept away or to exhaust their vitality in the attempt to grow above the water level. If, on the other hand, the floods drain off too rapidly, the stalk collapses for want of proper support and the ears are injured by falling in the water. Similarly, too much rain washes away the young jute plants, while over-inundation makes the fibre coarse and rooty.

Before 1947 Muslims from Bengal were able to take up land in Assam, but that outlet is now closed. While family limitation is the only permanent solution for this population problem, better cultivation might bring temporary relief. By restricting the jute acreage, more land might be planted with rice, but there are severe limits to that remedy because jute is the most valuable export of all Pakistan. Three quarters of the farmland carries only one crop a year. That crop, usually rice or jute, occupies the land for the seven to eight months over the wet season. For the rest of the year this land is bare, the soil baked hard by the sun. Double-cropping could best be encouraged by irrigation, which would enable *rabi* crops such as wheat, beans and vegetables to grow in the dry season. But big irrigation schemes are impracticable on account of the annual flood and the shifting rivers. Local schemes, depending on electric pumps for raising river water to nearby fields, are a possibility. There is one major scheme, the Ganges/Kobadak, in which canals in the western (un-inundated) areas south of the Padma will be fed by a spillway from that river; about two million acres will be watered.

In the hills around Chittagong maize, hill rice, vegetables and cotton are raised by shifting cultivators. This system is known locally as *jhuming* —after *jhum*, the name for the clearings. Along with their own crops the cultivators plant teak saplings supplied by the forestry authorities. By the time a *jhum* is abandoned after three years or so the new teak is sufficiently established to ensure the regeneration of timber.

Industry. About 300 factories employing 100,000 workers was all the industrial development that East Pakistan inherited in 1947—very small for a province of 40 million people. There were 10 cotton mills, a cement works on the Tertiary limestones near Sylhet, scattered jute-baling plants, tea and sugar factories. Progress is hindered by lack of fuel and minerals. The lignite of Sylhet is not worth mining. The substantial reserves of natural gas near Sylhet will help if a pipeline can be built as far as Dacca. Meanwhile, the search for oil continues. The forests of the Sundarban and the Chittagong Hills could be exploited further; bamboo is already used in the newsprint factory near Chittagong, and there are many small timber-based industries such as match-making, boatbuilding and furni-ture-making. A hydro-electric scheme on the Karnafuli river 27 miles above Chittagong provides power for the port and Dacca. Jute mills have been established at Dacca, Chittagong and Khulna. But the general prospects of sufficient industry to relieve rural poverty are not good.

Towns and communications. Dacca (population 600,000), the capital, is the largest town. Chittagong, the only port with wharves at which large ships can tie up, suffers the disadvantage of a long single-track metre-gauge link with the more productive parts of the province. A second port has been opened at Chalna, but there the ships lie in the river and unload into lighters. The smaller inland towns are administrative centres, steamer stations and collecting points for agricultural products.

The physical geography is against railways and roads. Many of the rivers are too wide to bridge even if their courses were stable. The great asset is the 3,000 miles of waterway, mostly between Dacca and the coast but extending to Sylhet and along the Brahmaputra and Ganges, on which there are regular services by large steamers. In addition, many other streams are navigable for part of the year. The steamer station at Goalundo is an instance of the difficulties of railway maintenance. There the Padma channel changes each year and the permanent way has to be relaid annually to whichever points are most convenient for steamers to tie up.

KASHMIR

Kashmir is a block of country almost 400 miles square. Its structure and relief are complicated. Six main divisions can be made out (Figure 28):

1—the Karakoram range with 33 peaks over 24,000 ft.;
2—the Indus furrow falling from 14,000 ft. in the south-east to 4,000 ft. in the north-west;
3—the ranges of the Himalayas, extending south-east from Nanga Parbat;
4—the Vale of Kashmir, an alluvial basin over 80 miles long and 25 miles wide, centred on the town of Srinagar (5,200 ft.);
5—the Pir Panjal range, rising to 15,000 ft.;
6—the Punjab borderlands, which include hilly country in the loose Siwalik sandstones, and a narrow strip of Punjab plain.

Atlas maps emphasize the strong relief of the parallel ranges, but in reality there are also extensive rolling plateaux with poor grass cover between 13,000 ft.–16,000 ft., especially on the northern flanks of the Himalayas and on the borders with Tibet. Only about four per cent of Kashmir is cultivated, and that mostly in the Vale where irrigated fields carry a great variety of crops, including rice, barley, wheat, oilseeds, tobacco and fruit. Population density per unit of *cultivable* land in the Vale is so great that in the lakes at Srinagar vegetables are raised on floating beds formed of reeds and manure. Rural densities are also fairly

(*Photo—Press Information Bureau, Government of India*)

9. Part of the Himalayan frontier of India: Kanchenjunga, and the jungly ranges of Sikkim and Nepal

10. Difficulties of road building in the Himalayas: a landslip in a gorge on the Kalimpong–Lhasa trade route

high on the Punjab borderlands, despite much forested hill country and irrigation difficulties associated with deep water tables and torrents in the Siwalik sands. Elsewhere, even north of the Himalayas at 13,000 ft., grain

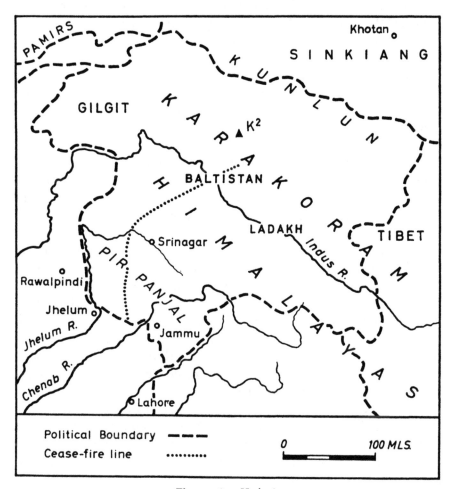

Figure 28. *Kashmir*

crops are grown on gravel terraces. Rice is planted up to about 7,000 ft. and barley at higher levels. Sheep raised for wool are kept on farms at lower altitudes in winter and driven to higher grazing in summer. North of the Himalayas, in Gilgit, Baltistan and Ladakh, physical conditions are very harsh and overall densities of population are below five to the square mile, but even there the pressure on the small patches of cultivable land is great. These northern areas once benefited from the trade between central Asian oases, such as Kashgar and Khotan, and the bazaars of the Indo-Gangetic lowland. For political reasons the Chinese have now discouraged this trade, and there is no longer much demand for pack-mules and yaks on the routes over the high passes of the Himalayas and Karakoram.

Two important sources of income are the tourists who visit the Vale and the craft industries based on silk and wool which have a world reputation for high quality. Timber and sericulture are occupations which provide exports.

The Kashmir dispute. Over three quarters of the total population of about five million are Muslim; less than a quarter are Hindus. The 40,000 Ladakhis near the border with Tibet are Buddhists. Only in the south-east is there a Hindu majority. Life would have been hard enough for all these people even under benevolent rulers. It was made harder still for the Muslim majority by the extortions of a succession of Hindu rulers who held power from 1846 until the departure of the British. In 1947 there were Muslim riots against the Hindu administration; Pathans from the north-west who came in to help the Muslims looted indiscriminately. The new government at Delhi sent Indian troops to oppose the Pathans, and Pakistan sent forces to stop the Indian advance. Since 1948, when a truce was arranged along the cease-fire line shown in Figure 28, both India and Pakistan have maintained strong forces in their respective areas of occupation. Both sides claim the whole state. The Indians in their zone have arranged social services and land reforms, and an assured market for the craft products of the almost wholly Muslim population of the Vale. The track from the railhead at Jammu over the Banihal Pass to Srinagar has been converted to a motor road at great effort (previously the only good route to Srinagar was by road from Rawalpindi in Pakistan). For their part, the Pakistan government have undertaken social welfare, road-building and recruiting local frontier-patrol forces in the wild country of Gilgit and Baltistan. Air-supply services have been a major achievement. Such airfields as exist are surrounded by steep mountain walls, demanding great skill from pilots in the bad weather conditions which are usual.

Permanent partition, probably along the present cease-fire line, seems the only practicable solution to the dispute. The Pakistani case is based primarily on defence problems, should all Kashmir fall to India, and on their need for control of the headwaters of the rivers that irrigate the Punjab. This second argument is not strong because the Indus, with which the Indian government would not interfere because of its distance from the Ganges plains, carries more water to Pakistan than all the Punjab tributaries combined. Pakistan, short of timber and minerals other than salt, has a sound claim to the forest resources and could make good use of the large bauxite reserves and small anthracite and iron-ore resources (as yet unworked) of the hills of the Punjab borderlands.

The need for Indo-Pakistani co-operation in this frontier area is underlined by the construction by the Chinese of a 100-mile stretch of their Tibet-Sinkiang highway through the north-east corner of Ladakh in the Indian zone without interference from the Delhi authorities.

Additional note on Pakistan (October 1963). At the outset in 1947 West Pakistan had the advantage of good railways and irrigation works, and a well-equipped port. Capital invested there was certain to yield quicker returns than if it were used in East Pakistan where communications are poor and construction work is severely hindered throughout the long wet season. Government policy has directed most of the foreign aid (as well as profits earned by East Pakistan tea and jute exports) into improving West Pakistan, with the result that the gap in living standards between the eastern and western territories noted on page 128 is widening. East Pakistan benefits little from the comparative prosperity of West Pakistan. Distance and the language differences between the two areas prevent migration of labour from the overcrowded east to the better opportunities in the west. These facts, together with the necessity of appointing people from the western area to senior posts in East Pakistan because of the lack of suitably qualified Bengalis, are a source of growing political tension. Many Bengali politicians accuse the central government of treating East Pakistan as a colony, and recommend political independence for the eastern territory. Recently the central government has responded to this situation by allocating a larger share of the available national funds to help East Pakistan.

12

Ceylon

Ceylon has a remarkably diverse physical geography, a fascinating history of human efforts to come to terms with the physical conditions, and a range of modern economic and social problems which also occur widely in other parts of Monsoon Asia. It therefore merits a longer chapter than its small area and population alone might justify.

The south-west quadrant of the island, which includes most of the hill country, has rain throughout the year and is known as the Wet Zone. The remainder of the island, the Dry Zone, has quite a high annual rainfall in parts but is everywhere subject to droughts—especially during the summer monsoon. Today the majority of the population of ten million live in the Wet Zone, where paddy farming on peasant holdings and the raising of tea, rubber and coconuts on plantations are the main occupations. For centuries the Dry Zone was malarial and very thinly populated: large tracts were, and still are, uninhabited. Yet a thousand years ago the Wet Zone was a tropical rain forest with hardly any people. At that time the Dry Zone was well populated; there were large irrigation tanks and paddy cultivation flourished well enough to support at least two capital towns the ruins of which are impressive.

The decline of the Dry Zone and the concentration of population in the Wet Zone still await a full explanation. Probably malaria, not endemic in the Dry Zone in early times, became a fatal menace when irrigation works decayed and provided the shallow, stagnant water ideal for malarial mosquitoes. The decay of the irrigation works might have been a consequence of a period of weak or corrupt government, or the Tamil invasions from south India, or both.

In recent decades the pressure of population on land in the Wet Zone and the need to reduce the dependence of the island as a whole on imported food have drawn attention to the possibilities of reclaiming the

Dry Zone. Early attempts at restoration of the old irrigation works and the large-scale planting of rice were unsuccessful, largely because malaria killed many labourers. Peasants from the Wet Zone could not be persuaded to take up the land. Since 1944 malaria has been effectively reduced by DDT spraying, and an imaginative programme of colonization, based on new irrigation works and the repair of old tanks and channels, has shown promising results. So many Wet Zone peasants are now eager to go to the Dry Zone that selection from the many applicants of those most likely to do well is an administrative difficulty. But it is a mistake to regard this recent work in the Dry Zone as an adequate solution to the food and population problems of Ceylon. The irrigation potential is very limited.

STRUCTURE AND RELIEF

About 270 miles from north to south and no more than 150 miles across at its widest, the island is built of pre-Cambrian schists, gneisses, quartzites and crystalline limestones similar to the rocks of peninsular India and folded into a synclinorium with an axis running north-east to south-west. Tertiary limestones, almost horizontal, occupy the surface in the Jaffna peninsula and extend for 90 miles southward along the west coast. The coal-bearing Gondwana sediments, which attain considerable thickness in India, are represented only by two insignificant patches devoid of coal seams. Remnants of three upland peneplain surfaces separated by escarpments have been recognized. The lowest surface, at a general level of 400 ft., embraces the hill-tops of the coastal lowlands. It has been greatly dissected in the west and south-west but remains prominent throughout the north and east. The second surface lies between 1,700 ft.–2,000 ft., and the third between 4,000 ft.–6,000 ft. Some geologists think the three surfaces represent three successive stages of uplift, the lowest surface being the one most recently raised above sea level. Others, impressed by the advanced erosion of the 400 ft. level and the comparative youth of the higher surfaces, suggest that the lowest is the oldest and that the middle and upper ones are the result of two block uplifts. The hill country reaches 7,360 ft. in Adam's Peak and exceeds 8,200 ft. on the mountain of Pedurutalagala. Relief is spectacular in the south-west, where the strike of the rocks along the southern closure of the synclinorium is roughly parallel to the coast: there are deep valleys in the crystalline limestones and the quartzites stand out as ridges.

Of the rivers, the Mahaweli Ganga is the longest and has the largest catchment; from headwaters in the hills south of Kandy it flows northward about 160 miles to enter the sea near Trincomalee. Other rivers are all much shorter. Alluvial lowlands are less extensive than the area between the coast and the 100 ft. contour on Figure 29 might lead one to

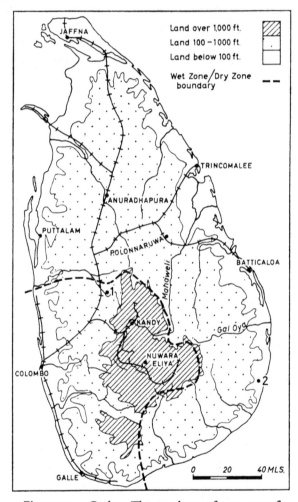

Figure 29. *Ceylon* The numbers refer to two of the stations in Table 6: 1—Batalagodawewa Tank; 2—Naula Tank

expect; this is of great importance in assessing the agricultural opportunities. Most of the alluvium is in the coastal zone below 100 ft., but even in
that zone much ground is taken up by numerous low pre-Cambrian
ridges which extend seaward to form rocky headlands. The alluvium is in
small basins between the ridges. Palm-fringed lagoons, some of them very
large, are characteristic of the east and west coasts; that at Puttalam, 70
miles north of Colombo, is 40 miles long. Apart from Trincomalee, there
are no good natural harbours.

MINERALS

The enormous reserves of iron ore associated with the pre-Cambrian
rocks of India have no significant counterpart in Ceylon. In the south-
west nine scattered deposits of medium grade are thought to contain
about 1,500,000 tons, enough to last only 35 years if the plans for the
small iron and steel works near Colombo were to mature. The ore could
be easily quarried. Coal would be a problem, but the steel furnaces would
be heated by hydro-electricity. Limestone for cement is plentiful. Plumbago (graphite), for which there is a constant demand in electrical industries, occurs in veins in the south-west. In value, the annual export of
plumbago is a mere 0·5 per cent that of tea. An interesting small-scale
industry is the search of secondary deposits for gemstones which derive
from the pre-Cambrian rocks. Sapphires, rubies and zircons freed during
the weathering of the old rocks find their way into mountain streams and
get caught up in the surfaces of small patches of alluvial clay. These clays,
easily located beneath later thin deposits of gravel and sand, are worked
with simple hand utensils to separate the valuable stones.

CLIMATE

The transition from Wet Zone to Dry Zone, as indicated by the vegetation, agriculture and density of population, is remarkably sudden. The
term Wet Zone is an adequate label for the area which receives rain
throughout the year: averages of some of the wetter stations exceed
200 in.; few are below 70 in. But Dry Zone is not an entirely suitable
description for the remainder of the island. Averages in much of the east

between the hills and Batticaloa reach 80 in.; they fall to around 50 in. over much of the north, and to 30 in. on the north-west and south-east coasts. Though subject to long droughts, the Dry Zone is not as arid as the dry areas of Pakistan, or even Burma. Throughout the island the average rainfall figures mask a substantial variability from year to year. In the Wet Zone this variability is rarely disastrous; but in the Dry Zone it usually results in long droughts, and in most years the actual monthly figures are less than the monthly averages. The averages in the Dry Zone are maintained at a relatively high level by infrequent years of exceptionally high local rainfall.

The data for the five stations in Table 6 illustrate the conditions in 1938. At Colombo in the Wet Zone rainfall was below the average in each of eight months, and at 64·76 was 26 in. below the average for the whole year. At worst, such a short fall might have impaired the rice yields where irrigation water was lacking. Batalagodawewa Tank, near the northern edge of the Wet Zone, received a mere 54 in., with a severe shortage in the main rice-growing season of the winter monsoon. Here the short fall of 25 in. must have brought crop failure on unirrigated fields. At Puttalam in the Dry Zone, the low rainfall of the winter monsoon, following the long dry season, resulted in an annual figure of only 30 in., 15·9 below the average. There, even dry crops which might just have got by on the average of 45·9 in. would have been severely affected. At Naula Tank, a Dry Zone station in the south-east, there were five consecutive months without rain but an exceptionally heavy fall of 21 in. (18 more than the monthly average) in February lifted the year's total nearly 15 in. above the average to 60·15. Such a heavy fall in one month is of little use to farmers but might cause disastrous local floods. At Jaffna the aggregate for the year was slightly above the average, but the November rainfall was 10 in. below the average for that month.

At coastal and lowland stations maximum shade temperatures often exceed 90°F. while minima below 65°F. are infrequent. The relative humidity is always high. In the hills up to 5,000 ft., night temperatures between 50°F. and 55°F. bring relief after day temperatures only a little lower than those of the coast. At Nuwara Eliya (6,170 ft.), one of the highest meteorological stations in Ceylon, night temperatures often fall to 35°F. in December and there is occasional slight ground frost.

TABLE 6

CEYLON: RAINFALL IN 1938, WITH OFFSETS FROM THE AVERAGE FOR THE YEARS 1911–30

(inches)

	Alt. (ft.)	J	F	M	A	M	J	J	A	S	O	N	D	Year
Colombo	24	1·74	5·97	8·14	15·57	3·48	1·94	4·10	4·77	5·74	4·86	3·82	4·63	64·76
		−2·3	+3·8	+3·3	+6·7	−11·5	−7·0	−1·9	+2·1	−1·3	−8·4	−8·5	−1·0	−26·0
Batalagodawewa Tank	422	2·60	5·95	10·95	10·88	1·05	0·68	2·55	2·58	3·03	3·05	5·70	4·99	54·01
		−3·1	+4·6	+5·7	+3·8	−4·7	−6·0	−1·4	+0·1	−2·9	−10·8	−6·8	−3·1	−24·6
Puttalam	10	3·80	6·35	7·30	3·94	0·42	0·00	0·26	1·22	0·08	2·14	0·70	3·84	30·05
		+0·2	+5·3	+4·0	−1·0	−3·3	−1·7	−1·1	+1·0	−1·7	−5·9	−9·3	−2·4	−15·9
Naula Tank	(near coast)	8·54	21·09	8·31	1·41	0·00	0·00	0·00	0·00	0·00	4·20	2·74	13·86	60·15
		−1·6	+18·1	+5·2	−0·5	−1·7	−0·4	−0·9	−1·0	−1·1	+0·2	−5·5	+3·9	+14·7
Jaffna	14	1·19	4·38	3·66	2·99	3·46	0·36	2·61	0·58	0·52	13·63	5·62	13·06	52·06
		−3·2	+3·2	+1·8	+1·5	+1·8	0	+2·0	−0·5	−2·6	+4·0	−10·7	+3·6	+1·2

In 1961 the total population, estimated at 10,167,000, was made up as follows:

	%
Sinhalese	70
Indian Tamils	12
Ceylon (or Jaffna) Tamils	11
Moors	6·40
Burghers	0·52
Europeans and others	0·08
	100·00

The Sinhalese majority, Buddhist in religion, are mostly peasant farmers. They also include substantial land-owning, professional, administrative and commercial elements. Indian Tamils, the largest minority, are Hindu in religion. They began to arrive in the nineteenth century to work on the plantations, employment that failed to attract the Sinhalese, and many returned to India after a few years. That custom has now declined, and immigration from India has been discouraged. The present Indian Tamil population is an essential labour force on the plantations; its political status is periodically a cause of friction with India. The second large minority, the Ceylon or Jaffna Tamils, are the descendants of thirteenth-century invaders from India. They now constitute a compact Hindu, Tamil-speaking, thriving peasant society in the extreme north, with important outliers along the east coast. The Ceylon Tamils adjusted themselves to western influences more easily than did the Sinhalese, and thus secured a strong foothold in commercial and professional life throughout the island: a few are Catholics.

The Moors, of Arab descent, are Muslim traders in coastal towns and villages. Their forbears came centuries ago when the Arabs dominated the Indian Ocean trade (see page 68). The Burghers, mainly town-dwellers, are descendants of mixed marriages between the Asian people of Ceylon and the Portuguese and Dutch.

The Portuguese set up trading posts at Colombo and other coastal places early in the sixteenth century; they introduced Roman Catholicism, which still survives, and Portuguese surnames such as Fernando,

Perera and de Silva are common today among low-country Sinhalese. While the Portuguese held only coastal sites, the Dutch who displaced them extended their authority gradually inland, although the Sinhalese rulers of the highlands around Kandy remained virtually independent until they surrendered to the British in 1815. Jaffna, the last Portuguese base in Ceylon, fell to the Dutch in 1658. The Dutch ruled until 1795 when, weakened locally by the withdrawal of forces during the Napoleonic invasion of Holland, they gave way to the British. The island became a British Crown colony, and achieved dominion status in 1948.

The Europeans are largely British. Many are in banking and commerce at Colombo; others are plantation managers in the highlands.

THE WET ZONE

On the low-lying alluvium that extends between the pre-Cambrian ridges, and in some places forms a recognizable coastal plain, a densely settled and almost entirely Sinhalese rural population is engaged in paddy cultivation. The ridges—many only 20 ft. or so above the paddy, others making prominent features over 100 ft. high—provide the settlement sites and carry roads and railways. On the higher ones some forest, probably secondary, is still to be found, but most of the ground has been cleared. Villages on these restricted sites are usually compact and linear. Gardens of vegetables and fruit trees surround the houses. Coconut palms are grown on the ridges and on sandy coastal tracts. There are large coconut plantations managed by commercial companies, and small groves owned by villagers. In this lowland country two rice crops a year are raised—one harvested after the summer monsoon, the other after the winter monsoon. Some fields are harvested twice a year, some only once. In the highland areas there is also a large Sinhalese population, though much hilly ground is occupied by tea and rubber plantations. Narrow valley bottoms are carefully terraced for paddy, and some of the world's most spectacular flights of paddy terraces are seen on valley sides. Many of the terraced slopes are so steep that the retaining walls of earth seem almost as high as the fields are wide. Continuous terracing from a valley bottom upwards on steep slopes for 1,000 ft. is common. All the fields are flooded when the rice is growing; the soil is consequently waterlogged and apt to slip—the fields move very slowly but continuously downhill, the lowest ones falling into the stream while new terraces are built at the upper end. Throughout the Wet Zone rice would probably mature on rain-fed fields even in exceptional years of poor rainfall, but in practice better yields are obtained by irrigation, which is available almost everywhere. On large flat areas there are modern schemes. In mountain valleys streams have been diverted into contour leats from which water is released to the paddy fields below.

Plantation agriculture, the economic mainstay of the island, is almost though not entirely limited to the Wet Zone. The large imports of rice and other foods, and the maintenance of a higher living standard for the population as a whole than obtains in most other parts of Monsoon Asia, are based on the profits of this industry. Yet the history of the plantations clearly underlines the economic dangers of exclusive reliance on this one source of wealth. In about 1835 coffee-growing made great strides in the highlands. Soon there were over 350 plantations, each owned and super-vised by one or a few Englishmen who had little knowledge of tropical agriculture and worked on capital borrowed at high interest rates. Profits were good for a time, but in 1847 the import duty on Brazilian and Javanese coffee, which had protected the Ceylon coffee in the English market, was halved. The consequent lower profits in Ceylon were not enough to pay the interest on the borrowed money, and many planters had to sell up. The estates were bought cheaply by new planters. These new planters, unencumbered by expensive loans, did well for a time, but were ruined by coffee blight caused by the insect *Hemileia vastatrix*. The blight first appeared in 1868, but its full effects were not apparent until 10 years later. Production fell from 1,000,000 cwt. of coffee in 1870 to 56,000 cwt. in 1895. A few planters had enough funds to turn to cinchona —too many as it turned out, because overproduction of this source of quinine brought a 75 per cent fall in prices between 1880 and 1886.

Tea plantations made a later start, and rubber became popular as an estate crop only after 1900, when demands for tyres for motors provided a large and increasing market. In the 1920's world production of rubber was well in excess of demand, and profits were generally poor. An inter-national restriction scheme devised to stabilize prices was unsuccessful, because the Netherlands East Indies refused to co-operate.

Unlike the tea and rubber estates, which were mostly British enter-prises, the third plantation crop, the coconut, has always been mainly Sinhalese-owned. In addition to copra, the trees yield *coir* (the fibrous outer husk of the nut), which is woven into mats and rope and the large palm fronds which are plaited together to make screens which serve as walls in temporary dwellings.

Each of the plantation crops finds optimum conditions within fairly definite ranges of altitude. Though they are sometimes grown above and below those limits, most of the rubber is on slopes below 2,000 ft., most of the tea between 3,500 ft.–4,500 ft., while the best coconuts are grown in coastal districts. Other commercial crops are cocoa, cinnamon and citronella, but of the total exports of the island in 1961 tea accounted for 71 per cent and rubber about 16 per cent.

The bulk of the tea, rubber and coconut products comes from large estates, but some is contributed by the smallholdings of Sinhalese farmers. There are now practically no individual European planters owning and managing single estates. The plantations have been grouped into larger and more efficient units, each managed by powerful commercial companies. On this account, economic catastrophe arising from shortage of capital is less likely to occur, and world markets could probably absorb all the future production of the plantations, provided quality of the crops is high and the prices stable. But other aspects of this industry give grounds for concern. The influx of labour from India is much reduced, and manufacturing industry in India might in future provide more attractive employment for the Tamils. (On the other hand, in recent years the proportion of Sinhalese workers has increased.) A large labour force is essential, for the picking of tea and the tapping of rubber cannot be mechanized. The labour force cannot produce much of its own food, and as there is no local food surplus from peasant farms the island is dependent on imported grain. Normally the cheapest as well as the most palatable grain is the surplus rice from countries in South-east Asia, where political unrest might at any time interrupt the supply. Experience has shown that a crowd of estate Tamils anxious about its food supply is more likely to blame the estate management than politicians in Burma, often with ugly results. Much will depend on the attitude of the Ceylon government. The political parties favouring nationalization of the estates are powerful enough to gain office; they create uncertainty, which might result in European companies withholding further investment.

Towns and communications. Broad-gauge railway lines, one from Jaffna and the ferry across Palk Strait to India, one from Trincomalee, another from the plantation areas of the highlands and a fourth from coastal areas in the south-west, all converge on Colombo (population 426,000), the capital and largest town. At Colombo there is little to remind one of the Portuguese and Dutch periods. Part of the city centre is built on a low headland and is called the Fort, though none of the early defence works are to be seen. Old graveyards contain a few Dutch tombstones. The artificial harbour is magnificent, but even with the recent improvements it cannot cope with all the shipping. There is usually so much congestion that several ships are at anchor outside, waiting for vacant berths.

Other towns are all small. In the south-west Galle was important in the Dutch period. Its anchorage is poor and has not been developed for modern shipping, but the town is of great interest for the massive walls built by the Dutch. The imposing crest of the Dutch East India Company, carved in stone, stands over one of the gates. In the highlands Kandy is a centre of religious pilgrimage. Nearby are the University of Ceylon and the Botanical Gardens of Peradeniya, famous for early research in rubber and tea cultivation. Hatton and Dikoya are very small market centres which owe their existence to the tea estates. But none of the highland towns except Kandy became important centres of social life for the European planters. The English preferred to meet at clubs established in open country among the estates. Nuwara Eliya at 6,170 ft. is a hill resort.

Roads are good, especially in the highlands. They were made originally so that bullock carts could manage the slopes; even in the steepest country they are consequently well graded and suitable for motor traffic.

THE DRY ZONE

Before modern colonization schemes were established after World War II the only densely settled parts of the Dry Zone were the Jaffna area in the north and the Batticaloa neighbourhood on the east coast. The Tamils of the Jaffna area cultivate with great skill and care. Their paddy is mostly rain-fed, while the intensive garden culture of tobacco and vegetables is dependent on irrigation from wells. Millet is grown on higher, unirrigated fields. The coastal alluvium around Batticaloa ranks high among the best rice lands of Ceylon. In both these areas, and around Puttalam on the west coast, coconuts are an important crop. For the rest, most of the land was uninhabited forest and scrub. Irrigation tanks and channels, largely in ruins, are striking evidence of former settlement; Minneriya, the biggest tank, has a water surface six miles long and two miles across (Figure 30). Scores more tanks are several hundred yards across, and there are countless smaller ones. Many of the bunds are too

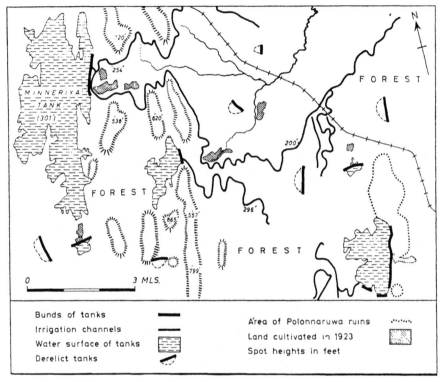

Figure 30. *Ceylon: Minneriya tank and old irrigation channels* Several of the channels have been restored, and much of this area is now cultivated as part of the Dry Zone colonization schemes

broken to impound much water, and the storage areas behind them are swamps. In this landscape of decay a few Sinhalese made a poor living, too weakened by malaria to restore the irrigation. Small patches of rice were grown in the once extensive paddy lands below the tanks. They also spent a good deal of time on dry crops such as millet, maize and cassava, grown in temporary hillside clearings, a form of shifting agriculture called *chena* cultivation, after the Sinhalese name for a jungle clearing. In a few instances where the *chenas* were near a road or railway even pineapples were grown for occasional export to the Wet Zone. Taken as a whole, the combination of *chena* and paddy was unsuccessful. The *chena* crops suffered greatly from droughts and required so much labour at the beginning of the wet season that the rice-planting was delayed.

Why did the Sinhalese who came to the island about the sixth century B.C. settle in the Dry Zone and neglect the Wet Zone? The answer appears to be that they may have come from the dry north-western parts of India, and found in the Dry Zone an environment in which their agricultural methods could be applied more easily than in the Wet Zone. Also the Dry Zone forest, less dense than that of the Wet Zone, was probably easier to clear.

How many people lived in the Dry Zone at the height of its prosperity during the tenth and eleventh centuries A.D.? The question cannot be answered. One could measure the area probably irrigated by each tank and estimate the number of people required to do the cultivation, but such an estimate would almost certainly be too high, for it is unlikely that all the irrigation works were operating simultaneously. More probably, new works were built as old and silted tanks were abandoned, so that only a proportion were productive at any one time. Unfortunately, the tanks cannot be dated by archaeological methods. Mr B. H. Farmer considers that the various estimates ranging from 5 to 17 million for the whole island are 'apt to be too large rather than too small'. Whatever the correct figure, the high cultural achievement of the Dry Zone civilization is proved beyond doubt by the ruins of Anuradhapura and Polonnaruwa. Rural food surplus was sufficient to maintain these two cities. The urban population, though perhaps not large, included the members of the court, the craftsmen and labourers who built the temples and the sculptors who made the enormous granite images of Buddha.

The modern agricultural colonies. Faced with acute overcrowding in the Wet Zone and the need for more home-produced food, the authorities in Ceylon regard the emptiness of the Dry Zone as a challenge. Attempts by private enterprise shortly after World War I to repair irrigation works for the large-scale production of rice all ended in failure. The reasons were malaria, flood damage and lack of capital. The Minneriya Development Company, which sought to open up 9,000 acres for the mechanized cultivation of paddy, came to nothing. Government schemes which were

11. The Great Wall of China, north of Peking

12. Hong Kong: Kai Tak airport

pressed ahead after 1944 had the advantage of effective control of malaria by DDT spraying, and several large colonies were established, the most ambitious being in the Gal Oya drainage area in the south-east. An earth dam 150 ft. high has been built to impound a reservoir lake extending 10 miles upstream. The scheme is expected to bring irrigation to 72,000 acres of scrub, to improve the water supply to existing irrigation channels of a large coastal area and to provide 100,000 acres of unirrigated land for dry crops. All this, together with small industries powered by the 5,000 kW hydro-electric plant at the dam, might ultimately support about 250,000 colonists from the Wet Zone.

Among the scattered groups of farmers away from the colonization schemes life is much as it was in the 1930's, but with malaria less of a menace and some hope of state help with agricultural improvements.

Local enthusiasm in political circles has bred extravagant claims that the Dry Zone can be developed sufficiently to solve the island's food and population difficulties. But a sober assessment of the physical and economic geography indicates that it would be optimistic to expect the colonization schemes to absorb even the annual natural increment of Wet Zone population, and that only for a few years. There can be no question of providing sufficient land and irrigation water to reduce the existing density of Wet Zone population, though a rice surplus in the Dry Zone could substantially reduce the need for imported food. The only solution to the population problem is reduction in family sizes. The considerations which lead to these conclusions are:

1—The impervious pre-Cambrian rocks, the foundation of most of the Dry Zone, are poor aquafers. The rivers are the only considerable source of irrigation water. (The heavily populated Jaffna peninsula is an exception. There the limestone contains abundant water)

2—Most of the Dry Zone rivers have only small parts (if any) of their catchments within the Wet Zone. Consequently they are reduced to chains of pools in the dry season. The total amount of river water available is severely limited.

3—There is not a lot of good alluvial lowland in the Dry Zone. Diversion of water from one catchment to another across the pre-Cambrian ridges so as to concentrate irrigation on the best soils would be expensive.

4—The irrigation tanks are necessarily shallow: the surface area is large in relation to the volume of water. Losses by evaporation are therefore great.

5—Though the Wet Zone is crowded and many people are very poor, living standards in general are higher than in most other parts of Monsoon Asia. These higher standards are ultimately provided by the profits of the plantations. Life in the Wet Zone is relatively easy. People will not willingly move to the Dry Zone, to years of hard labour building their own irrigation works and clearing scrub. They are prepared to go only when the government pays for all this preparatory work and provides dwellings. The cost is high, about £1,500 a family.

The largest of the Dry Zone towns, Jaffna, has a population of about 80,000. Trincomalee has a good anchorage and is on the railway system, but its distance from the main centres of production has hindered development since the British period, when it was important as a naval base and Royal Air Force station.

CEYLON: THE FUTURE

The principal economic challenge is how to support a rapidly increasing population. In 1957 there were 9,165,000 people. By 1975 there will probably be 12,500,000. There is no certainty that the plantation products will continue to command good prices in oversea markets so as to pay for big food imports indefinitely. Prospects for industrialization are slight. People are employed in estate factories which process rubber and tea; a small textile and plywood industry has been established; and there is a cement works in the limestone area of the north. Colombo has a range of small industries, and also engineering works connected with the railway and the port. More hydro-electricity is becoming available from schemes in the highlands. But all these things are on too small a scale to have much effect on the population issue. Improvement of agriculture, better use of the Dry Zone, and above all the prevention of population growth beyond a reasonable level, are the only sound policies for the future.

Limited as the economic prospects are, in 1948, when independence was conferred, they were better than those of most countries in Monsoon

Asia. The new state inherited two valuable assets—a flourishing plantation industry and a tradition of good race relations between the Sinhalese majority and the Tamil minority. The Sinhalese politicians disastrously wasted their opportunities by encouraging racial conflict: in 1958 their Official Language Act prescribed Sinhala as the sole official language of Ceylon. Despite provision for what was called 'the reasonable use of Tamil' in law courts and education, Tamil distrust of the Sinhalese majority grew. By 1961 the situation was ugly enough for a state of emergency to be proclaimed. The Indian and Ceylon Tamils, formerly distinct groups with little social or political intercourse, have come to regard the Sinhalese as their common adversary, and the possibilities are now alarming. The united Tamils would form a substantial minority of over two million, and there are 30 million Tamils as potential support in south India only 20 miles across Palk Strait. A strike of Tamil labourers on the estates could result in ruining the country.

13

China

Some aspects of the historical geography of China up to the end of the Manchu dynasty are given in Chapter 6 (pages 63–7). There was a succession of empires. Each in its turn matured to a climax of power, administrative efficiency and territorial extent, later to fall apart in the confusion, rebellion or invasion which was the direct or indirect result of oppressive and incompetent government. This pattern continues into recent times. The Nationalist régime, which replaced in 1911 the corruption and inefficiency of the closing decades of Manchu rule, promised well at the start under the leadership of Sun Yat Sen, only to become as corrupt as its predecessors by World War II. After a civil war it was replaced on the mainland in 1949 by the Communist régime, though its leader, General Chiang Kai-shek, withdrew what was left of his forces together with about a million refugees to the island of Taiwan (Formosa). There Chiang Kai-shek continued to claim authority over all China. The situation was absurd: the strong and progressive Communist régime, ruling effectively from Peking over more than 500 million people, was denied full international recognition; the remnant of the Nationalists, governing only the eight million people of Taiwan, was allowed to retain its seat on the Security Council of United Nations, as though it had the authority to speak for all the Chinese people.

China Proper and Outer China. Throughout the long and complicated history since the Han empire, China can be regarded geographically as falling into two divisions: (a) the densely settled areas south of the Great Wall and east of the Tibetan plateau, and (b) the uninhabited and sparsely populated outlying areas to the north and west over which the Chinese had control. The first division, characteristically Chinese in its society and culture, includes the large and crowded lowlands of the Hwang Ho,

Yangtze and Sikiang together with the neighbouring and intervening uplands which are also closely settled except in their higher parts. It became known in western literature as China Proper. The other division, for which Outer China is as good a label as any, is very different. It embraces the plateau of Tibet and the deserts and grasslands of central Asia as far as the Pamirs and the borders of Siberia. Throughout this area there are very few Chinese except in Manchuria, to which they have emigrated in great numbers during the last 60 years. On the grasslands north of the Great Wall are the Mongol pastoralists. There are other pastoralists farther west on the grassland borders of the Taklamakan Desert and in Tibet. In the oases of the Taklamakan area, where the population density approaches that of lowland China, 90 per cent of the farmers are Uighers, Muslim people of Turkic speech. Only 10 per cent are Chinese. The farmers thinly scattered over south-eastern Tibet are also different from the Chinese in language and religion: they are Lamaist Buddhists and speak Tibetan.

During the Manchu period China Proper consisted of 18 provinces, while the whole of Outer China was considered subject territory of lower political status. Of all Outer China only Sinkiang was raised to provincial status, and that not until 1878. In this political sense the distinction between China Proper and Outer China is no longer valid, for several provinces have been created in the outer area. But the two divisions are worth retaining because they remain well grounded in the geographical sense. China Proper, with an area of roughly 1·5 million square miles, has a population of over 500 million. Outer China, with double that area, has a mere 50 million people and over half of those are in Manchuria.

Another contrast between the two divisions is presented by the pattern of railways in Figures 33 and 34 (pages 172–3). A rail network serves China Proper, and parts of Manchuria and the southern fringe of the Mongolian grassland. In the remainder of Outer China there were no railways until the 1950's when the Communists began work on two lines leading to the U.S.S.R.—one through Sinkiang, the other northward across the Gobi Desert. A third railway, from Lanchow westward and then southward to Lhasa, is projected. But the Chinese have always appreciated the need for good communications to link outlying parts of their empire with administrative centres in the east. This fact is illustrated by the published intentions of the Chinese government in 1921. They contemplated for Outer China a network of railways only a few of which

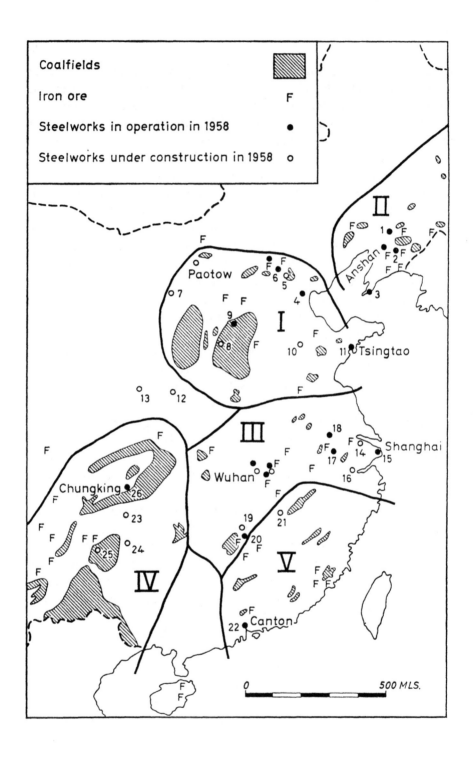

Coalfields

Iron ore   F

Steelworks in operation in 1958   ●

Steelworks under construction in 1958   ○

II

F   1   F   F

Anshan   F 2   F   F

F   F

3

F

○

Paotow   F   F

6   5

○ 7   F   F   4

9   I

○ 8   F

10   11 Tsingtao

○ 13   ○ 12   F

III

F   18

F   F   F   Shanghai

F   17   14   15

F   Wuhan   F   F

F   16

Chungking ● 26

○ 23   19   ○ 21

F   F F   ○ 24   F   20   F

F   ○ 25   F

F   IV   V   F

F   F F

F

F   Canton

F   22

0   500 MLS.

are shown on Figures 33 and 34; this economic folly involved twelve separate lines crossing the Gobi Desert, several lines linking the Sinkiang oases and seven lines converging on Lhasa. Today the Chinese are engaged on road construction to link all Outer China with railheads in China Proper. That work is soundly based on strategic and economic considerations.

POWER AND INDUSTRY

Coal reserves. Alone among the countries of Monsoon Asia, China has very large reserves of coal. The proved deposits amount to about 10 per cent of those of U.S.A., and this is a conservative estimate. Additional reserves will probably be discovered as geological work continues, especially in Outer China.

Coal is found in almost every province. The main fields, which are shown in Figure 31, fall into six groups:

1—*North China.* The two enormous fields of Shensi and Shansi, which together contain some 75 per cent of the total known reserves of China, are here. The four fields in a belt running westward between Peking and Paotow are especially valuable on account of their coking coal. There are several small fields in Shantung.

2—*Manchuria.* In the Fushun field of Manchuria the world's thickest seam of bituminous coal is quarried in surface workings. This seam has a maximum thickness of nearly 400 ft. and dips at about 30 degrees. Other fields, notably that at Penchihu, have good coking coal.

3—*The Yangtze valley* has many small fields, some with coking coal.

4—*The south-west.* The Szechwan coalfields have exceptionally good coking coals but the seams are thin. There are larger reserves to the south in Yunnan and Kweichow.

Figure 31. *China: coalfields, and the iron and steel industry* The steelworks indicated by numbers are: 1—Mukden; 2—Penchi; 3—Dairen; 4—Tientsin; 5—Peking; 6—Shihkingshan; 7—Shihtsuishan; 8—Linfen; 9—Taiyuan; 10—Tsinan; 11—Tsingtao; 12—Sian; 13—Paochi; 14—Soochow; 15—Shanghai; 16—Hangchow; 17—Maanshan; 18—Nanking; 19—Siangtan; 20—Pingsiang; 21—Nancheng; 22—Canton; 23—Tsunyi; 24—Kweichow; 25—Kaili; 26—Chungking. (This diagram is based partly on 'China and her race for steel production' by Brian Crozier, published in *Steel Review*, July 1959)

5—*The south-east.* Here the reserves are very poor and of low grade.

6—*The north-west.* No maps showing coalfields of this area are available. It is thought that the several fields of the province of Sinkiang contain 13 per cent of the Chinese reserves. There is a 60 ft. seam near Tihwa, and several fields on both flanks of the Tien Shan range.

In addition, there are small reserves in Taiwan.

Coal production. Of the 120 million tons raised in 1957, the Shensi and Shansi deposits, which account for practically 75 per cent of the entire Chinese reserves, contributed only 11 per cent. The northern group of coalfields as a whole contributed 55 per cent, but most of this came from the smaller fields served by railways. Manchuria produced 36 per cent. About 5 per cent was raised in the south-west group, and a mere 2·5 per cent in the Yangtze valley. The south-east and north-west groups contributed very little. Thus Manchuria and north China together account for over 90 per cent of the total output. The total coal production for 1960 was reported to be 450 million tons.

Oil resources are modest. A belt of country about 700 miles long at the foot of the Nan Shan in Kansu is known to be oil-bearing, though production so far is limited to the district around Yumen (Figure 33). There is a small output but probably useful reserves in Sinkiang. Sinkiang and Yumen are respectively 1,700 and 1,200 miles from the nearest large Chinese markets, and it would probably be cheaper at present world prices to import foreign oil, should a big demand arise. Wells in Szechwan have yielded small quantities of oil for centuries, but no big reserves have been found. The oil shales which overlie the coal at Fushun in Manchuria have long been an important source of oil fuel. Exploitation is cheap on account of the location and good railway communication; costs are further reduced because the shale has to be removed before the coal can be worked.

Hydro-electric potentials are enormous, especially in the wetter areas of the south in the Yangtze and Sikiang drainage. The cost of dams is such that thermal electricity is cheaper wherever coal is available within 300 miles, therefore the prospects of hydro-electric development are uncertain. During the Japanese occupation of Manchuria in the 1930's large works were completed on the Sungari and Yalu rivers. Politicians throughout Monsoon Asia have a weakness for multi-purpose river-control works, however expensive. The Chinese have a scheme for 41 dams on the Hwang Ho with a total generating capacity of 23 million kilowatts, nearly 10 times the entire Chinese generating capacity from all sources of energy in 1954. A spectacular start was made with a dam 500 ft. high at the Sanmen Gorge, some 40 miles below the confluence with the Wei river.

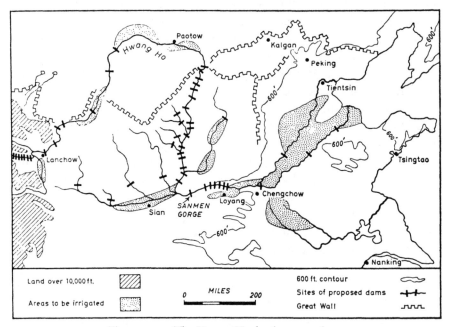

Figure 32. *The Hwang Ho development scheme*

The whole scheme, which will take decades to complete, might be justified in terms of flood control and irrigation on the northern plain, but it is not a cheap source of power (Figure 32).

The iron and steel industry. Deposits of iron ore are widely scattered (Figure 31). Modest in amount for so large a country, they are sufficient for the foreseeable needs. It might turn out to be cheaper to import ore from the Philippines or Malaya than to work the more remote fields of the interior. Two thirds of the reserves, mostly of low grade, are in Manchuria. Higher-grade ore is scattered in smaller deposits, notably north-west of Peking, near Paotow, in the Yangtze valley and on the island of Hainan. The world's largest reserve of tungsten (essential for hardening steel) and some manganese are in the south-east in Kiangsi. Other minerals of which China has good stocks are tin (in the south-west in Yunnan and Kwangsi), antimony and copper (in Taiwan and Manchuria).

In the 1930's the iron and steel industry was dominated by the works at Anshan in Manchuria started by the Japanese in 1919. There were also about a dozen smaller works of minor importance. In 1946 the Anshan works, which had been extended by the Japanese to an annual capacity well over a million tons of steel, was almost completely dismantled by the Russian occupation forces. Much of the plant was sent to steel centres in the U.S.S.R. The Communist régime in China, faced with the need for rapid industrialization, secured Russian help in building a new Anshan which came into full production about 1954, and this development became part of a plan to raise the Chinese annual steel production to 29 million tons by 1967—40 million tons by 1972. Set against the British steel output of 20 million tons in 1957 these targets seem high, but in assessing them in terms of relative living standards one should remember that China has 10 times the population of Britain and would need a proportionately higher steel output if anything approaching western living standards is to be achieved. Anshan now produces about five million tons of steel a year, and the smaller works elsewhere have been reconstructed and enlarged.

There are two new major schemes for integrated iron and steel works on the scale of Anshan: one at Wuhan on the Yangtze, the other at Paotow in the north-west. But works of this size turn out to have three disadvantages: they are very expensive to build, the concentration of steel capacity in three major centres is a strategic weakness and the consumption of ore and fuel in enormous quantities at three centres calls for expensive transport from distant and scattered deposits. The plan was therefore altered halfway through. Development is to continue at Wuhan and Paotow, but priority has been given to 14 new and smaller works indicated in Figure 31. The smaller works cost less to build, they are well sited in relation to local ore and fuel and they have the strategic advantage of dispersion.

The significance attached by the Chinese to iron and steel production is so great that in 1958, when progress at the steel works failed to satisfy the government, peasants and townspeople were urged to set up small furnaces throughout the country to use local ore and scrap iron. Fortunately there had been a good harvest and the peasants had time to spare. Thousands of backyard furnaces made from oil-drums and local brick were set up, but the scheme was soon abandoned as wasteful. The 'steel' produced was little better than cast iron, and in many instances it could not be moved far from the furnaces because the roads (if any) were bad.

Other factory industries. Before World War II heavy-engineering and chemical industries were associated with the iron and steel plants, mainly in Manchuria. Lighter industries, such as textiles, paper, leather, food-processing and shoe-making, were concentrated in three areas: the industrial zone of Manchuria centred on Mukden and Anshan; a northern area which embraced Peking, Tientsin and Tsingtao; and the Shanghai-Nanking area. During World War II there was some industrial progress in Szechwan while Chungking was the national capital, but expansion of industries beyond those areas was hindered by poor communications. The railways are shown in Figures 33 and 34: even Manchuria and northern China are only moderately provided, and in the rest of the country there are vast areas without railways. Of the rivers, only the Yangtze up to Chungking is of more than local value for navigation. There were hardly any roads, though this is being improved; one can now motor from Peking to Canton, Tihwa and Lhasa, but road surfaces are mostly earth or gravel. A comprehensive range of industrial development is to be expected around the new steel centres. Already most towns of over 50,000 population have some factories, and there will be more as communications improve.

THE REGIONAL DIVISIONS OF CHINA PROPER

There are striking contrasts between the north and south. The north has long, cold winters, and a low annual rainfall limited to the period of the summer monsoon. Trees are sparse. The countryside is brown and dusty except in the wet season. The south, on the other hand, is protected from the full influence of the continental high pressure in winter by the Tsinling mountains: consequently, winters are milder than in the north. Rainfall is higher and more evenly distributed throughout the year. Trees are abundant. The countryside is always predominantly green. The transition between these northern and southern conditions takes place in a belt of

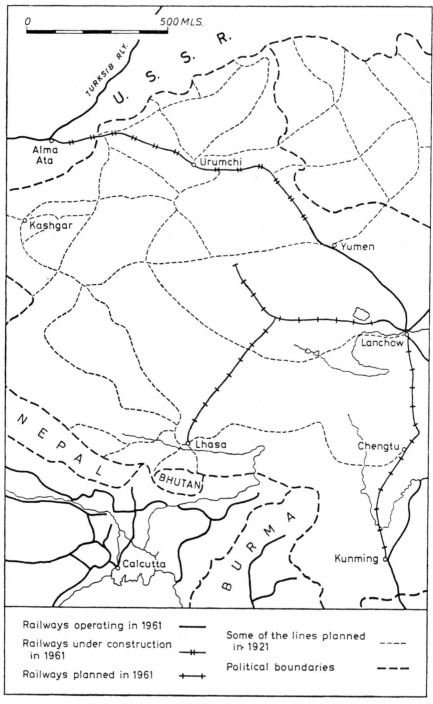

Legend:

Railways operating in 1961	———
Railways under construction in 1961	—++—
Railways planned in 1961	—+++—
Some of the lines planned in 1921	- - - -
Political boundaries	— — —

Figure 33. *China: railways* (1)

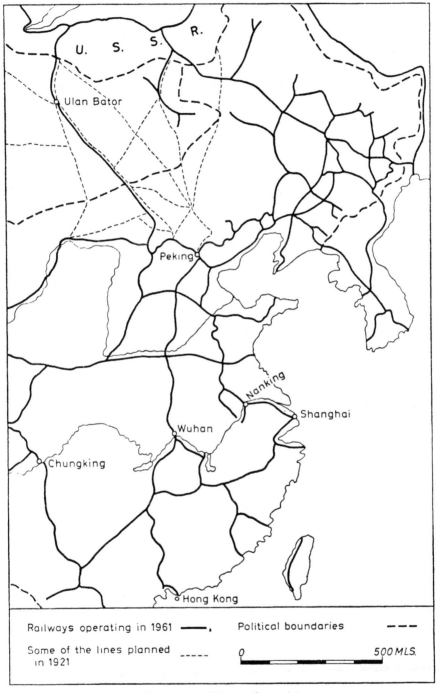

Figure 34. *China: railways* (2)

country extending from the western end of the Tsinling range to the sea north of Shanghai.

In the north farmers must contend with disastrous floods, low and uncertain rainfall and a growing season of only four to six months. Cold northerly winds, which often persist well into the spring, might delay planting or kill established crops. In four major famines between 1810 and 1850, 45 million people died. Ten million died in the famine of 1877. Tragedies on that scale no longer occur because communications now enable food to be brought to stricken areas from outside, but crop failures, resulting locally in prolonged hunger and malnutrition, are common. To set against these adverse physical conditions, the north has one major asset in its unleached soils, which include the loess of the highlands in the north-east (Figure 35) and the young alluvial soils of the plain, which themselves derive mainly from the loess.

The physical environment in the south is kinder. Fewer floods, a rainfall more secure and better distributed and a growing season of seven to twelve months make things easier for the farmer, though the rural population dependent on local food supplies is so large that shortages are common even there.

It is remarkable that although the opportunities for farming are better in the south, Chinese intensive agriculture originated in the harsher north (see pages 63, 64). The emergence in early times of powerful agricultural states in arid or comparatively unattractive country, while wetter areas now densely settled were neglected, is a recurrent theme in Monsoon Asia. In addition to this example in China, one finds it in the Dry Zone civilization of Ceylon and the Indus civilization.

China Proper is roughly 1,200 miles square. It can be broken down into divisions such as the Shantung highlands, the loess highlands, the North China plain and the Szechwan Basin. But divisions of this kind, based on physiography, have two disadvantages: they are too numerous to remember easily, and even when memorized they emphasize the scenic diversity of China Proper, but exclude the factors which impart geographical homogeneity to quite large tracts—sometimes regardless of variation in structure and relief. The soundness of this view can be appreciated from the pioneer work of Professor J. L. Buck on the agricultural regions of China. Using agricultural criteria alone, Buck discerned eight major regions. Since China is, and will probably remain for decades, a predominantly agricultural country despite the recent efforts at industrialization, these eight agricultural regions form a more significant basis for regional study—provided that one takes account of the variations in physical geography within each region.

Professor J. L. Buck's survey

The survey was conducted between 1929 and 1933. Research assistants collected statistical data on 16,786 farms in 170 representative localities. Information was gathered about crops, livestock, sizes of farms, irrigation, the proportion of the total land area under cultivation and the density of rural population, and the results were published in 1937. Buck claimed that they provided no more than a reliable reconnaissance survey, but after an interval of 30 years they remain the only detailed study embracing the whole area south of the Great Wall. The eight agricultural regions are shown in Figure 35. Their boundaries for the most part indicate zones of transition from one region to another, not abrupt changes. Much has happened in Chinese agriculture since the field work was done. In recent years collectivized agriculture has been progressively substituted for the long-established form in which the individual peasant worked as he thought best. But despite changes in social organization, population density, and perhaps also to a minor extent in crops, the eight regions of which some account is given below are still valid. While reading the following paragraphs, reference should be made to a good atlas map.

(1) THE SPRING-WHEAT AREA

In this northern belt of hilly country at a general level of 2,500 ft. to 3,000 ft., rising over 5,000 ft. throughout Kansu, there is very little flat land except along the Hwang Ho. Though the area includes the northern part of the loess highlands, conditions are poor for farming. The uncertain rainfall averages 14 in., about half of which falls in heavy showers in July and August. The growing season is little more than four months. Winters are extremely cold, with light snow. The crops—wheat, with smaller acreages of potatoes, barley, millet and oats—are all spring-sown. In spring the melting snow moistens the hard ground and makes ploughing easier. Often there is no spring rain, the crops get a late start and there is hardly time for them to mature before winter. There is not much irrigation except along the Hwang Ho. Excellent peaches, apricots and melons are grown.

Agriculturally, the spring-wheat area is transitional between the better farmlands immediately to the south and the dry grasslands and deserts of Mongolia. The Great Wall, built by the Chinese as a northern boundary to the country suitable for their form of intensive agriculture, lies almost entirely within it. At the time of Buck's survey less than 20 per cent of the total area was cultivated, though the figure for the part lying south of the Wall would be substantially higher. While the overall density of population is the lowest in China Proper, the pressure on the cultivable land is high. There is little hope of widespread improvement or extension of agriculture, though the government scheme for controlling the Hwang Ho might provide irrigation to parts adjacent to the river.

Of the towns, those north of the Wall are more appropriately dealt with under Inner Mongolia. South of the Wall, Lanchow is the largest, with over 500,000 people. Capital of Kansu and a long-established trading centre on the main route to Sinkiang, it has acquired additional importance in recent years as a railway town. There are engineering works, and an oil refinery fed from the Yumen oilfields 400 miles to the north-west. Towards the east, Tatung is a growing coalmining town with engineering and cement works.

The population of Kansu includes a 14 per cent minority of Muslim Chinese. Islam came to this area in the eighth century by the central Asian routes.

(2) THE WINTER-WHEAT/MILLET AREA

Throughout this area rainfall is slightly higher than in the spring-wheat country, and winters permit autumn-sown crops to survive. Winter wheat and cotton are the principal crops on the best land, with millet on the poorer hillsides. There are small acreages of maize, rice and kaoling in valley bottoms. Kaoling is a spring-sown sorghum, less resistant to drought than millet; it is fed to farm animals, and its stalks are used as small structural timber and for fuel. Temperate fruits do well, and there are even some vineyards. The Hwang Ho and its tributary the Wei Ho divide the highlands into three parts: the loess highland in the north-west, the highlands of the north-east and the Tsinling ranges of the south.

The loess highland, mainly in the province of Shensi, is plateau country between 2,000 ft.–3,500 ft. The loess, attaining a maximum thickness of

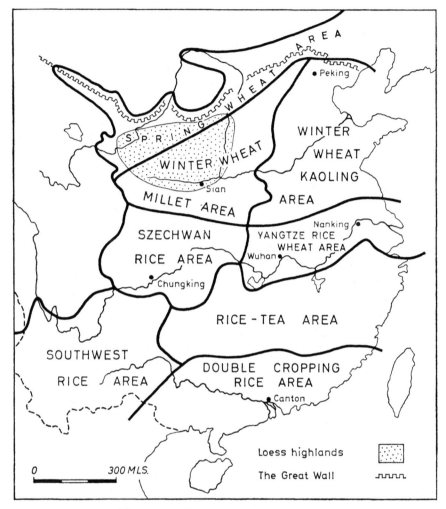

Figure 35. *China: agricultural regions*

at least 500 ft., is a porous, moisture-retaining material in which rivers
cut deep ravines. There are few trees and practically no natural woodland.
Headward erosion by the lateral feeders along the main streams has reduced the extent of the rolling upland surface appreciably even within
living memory, and terracing (to minimize erosion) has been practised
for centuries. In recent years conservation measures have been intensified.
Roads have been cut deep by centuries of cart transport: many are
countersunk as much as 40 ft. into the landscape and flanked by vertical

cliffs of loess. Though the loess is unconsolidated, its grain size, cleavage and chemical composition are such that the cliffs are normally quite stable. Caves are easily excavated; many people live in roadside caves beneath their fields. 'Chimneys' are cut through the overburden, so that domestic smoke may be seen rising from holes in the earth in the middle of a crop of wheat.

Earth tremors, to which the area is subject, cause disastrous landslides and 250,000 people died during the earthquake of 1920. In good years crop yields are satisfactory; they could be much higher if irrigation were available. By Chinese standards, the population density away from the rivers is not very high, but all the cultivable land is fully occupied.

There is not much loess either to the east of the Hwang Ho or south of the Wei Ho. The highlands to the north-east form plateau country between 2,000 ft.–6,000 ft. drained by the Fen Ho, with mountains on the eastern border reaching 10,000 ft. The Tsinling ranges of the south are rough country with farming only on lower slopes and valley bottoms.

The Hwang Ho valley in this area is a rocky gorge of little use for settlement or as a routeway. The plain of the Wei Ho, 200 miles long and 40 miles wide at a general level of 1,200 ft., and that of the Fen Ho, have irrigated land. Both are densely settled, and produce substantial quantities of cotton. Formerly, most of the cotton was sent to the mills of Shanghai, but the textile industry is now well established close at hand in Sian and other towns. Both the Wei Ho and Fen Ho valleys have railway links with the rest of China.

The winter-wheat/millet area contains the greater part of both the Shensi and Shansi coalfields. These two coalfields have great potentialities, though at present there are few large mines. Sian (population probably about one million), once the capital of the Han empire, is a growing industrial city. At Taiyuan there is heavy engineering and iron and steel works. Linfen and Paoki have emerged as steel producers under the Communist régime.

(3) THE WINTER-WHEAT/KAOLING AREA

Here the principal crops are winter wheat, kaoling, cotton, millet, barley, corn and soya beans. The wheat, planted in autumn, is ready for harvest early enough in the following year for a second crop of soya beans, corn or groundnuts to be raised on the same ground before winter. About a third of the land is double-cropped in this way. The cotton crop accounts for roughly two thirds of the entire Chinese production. Average rainfall decreases from about 35 in. in the south to 17 in. in the north. Some rice is seen in the south, but very little towards the north, where alkaline soils are unfavourable and the natural rainfall must be supplemented by irrigation. There are two main physical divisions: the North China plain and the mountainous Shantung peninsula. Borderlands on the west and north are hilly.

The North China plain (Figure 36) has a population of about 140 million, nearly a quarter of the entire population of China; though mainly built of sediment deposited by the Hwang Ho, hardly any of the plain now drains to that river. For centuries the Chinese have maintained dykes to confine the river to a permanent channel, with the result that silt which would under natural conditions be distributed widely over a flood plain accumulates in the river bed. Thus the river has raised its bed above the general level of the plain. At low water the river level is five to ten feet higher than the land outside the dykes; during floods it is often 25 ft. higher. Millions of people live below the level of flood water. Rural densities exceed 1,000 persons to the square mile over great areas. Both north and south of the Hwang Ho the plain is drained by minor channels to the sea, or occasionally to shallow lakes where the water evaporates.

The sedimentary burden washed down by the Hwang Ho, principally from the loess country, is so great that banks of silt deposited in the lower course by one spate of flood water often impede the flow of a subsequent flood following close behind. The water rises until it finds a weak place in the dykes and overflows them. Though they are protected in places at critical times by mats of kaoling straw, wide gaps are easily cut by flood water since the dykes are only made of earth (there are no roads or railways to bring concrete or building stone to the river banks). Minor inundations covering hundreds of square miles are frequent, and there have been many major disasters in which the river has broken away to establish itself in a new course. There have been about 15 separate courses to the sea, all radiating from the locality of the city of Kaifeng. In 602 B.C. the river changed from a northward course to the sea near Tientsin to one running out south of the Shantung peninsula. This southern course was re-established in 1324 after several northward alignments had each for a

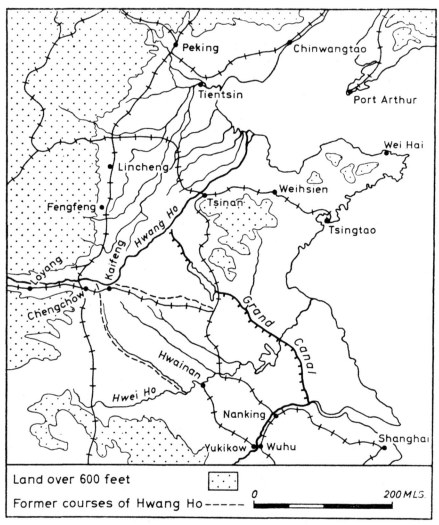

Figure 36. *The North China plain*

time been the main outlet. The present course to a mouth 250 miles north of the southernmost outlet was adopted after the flood of 1851. At each major change of course the devastation and loss of life were enormous. In 1938, when the Nationalist Chinese cut the dykes to divert the river to a southern course to hinder the invading Japanese, there were six million refugees and probably 500,000 were drowned. On that occasion most of the water spread southward to the Hwai river and ultimately reached the

Yangtze. The dykes were repaired in 1947 and the river restored to its former course past Tsinan.

The mountainous Shantung province, rising to 5,500 ft., is divided into two by a wide rift valley known as the Kiaolai corridor between Tsingtao and Weihsien. Once forested but long since cleared of useful timber, the hills have little to attract farmers. The small coastal alluvial areas and some valley floors including that of the rift are as densely settled as the North China plain.

In the winter-wheat/kaoling country as a whole, about 10 per cent of the crop acreage was irrigated at the time of Buck's survey, most of this irrigation coming from wells in the drier northern part. The proportion today is higher. The Kwanting reservoir north of Peking provides irrigation water along the lower course of the Yungting river. Major irrigation schemes would be a tremendous asset in this area of unreliable rainfall and vast population: this, with the threat of floods, is the urgent reason for more effective control of the Hwang Ho. Strengthening the dykes and the provision of emergency flood channels parallel to the river cannot be a lasting solution, since deposition is always bound to catch up to a point where high floods will overflow the banks. It has been suggested that flood water should be passed into settling basins on the plain. New basins would be formed as older ones filled with silt to a level at which farmers could settle on them without risk of flood; just such a scheme is impracticable because large areas would be taken out of cultivation and there is no other land for the population that would be displaced. There seems no alternative to the expensive long-term scheme referred to on page 169 for dams throughout the length of the river. The enthusiastic Chinese reports on this scheme are silent about the risk of reservoirs filling with silt soon after the dams are built. The scheme is not necessarily unsound on that account, but the problem of controlling sedimentation would be difficult. The Sanmen Dam is on the western edge of the area.

Communications and industry. Of all the country south of the Great Wall, this area has the best railway network. The Hwang Ho is almost useless for navigation except locally by small junks and other country craft. An industrial belt extends from Peking and Tientsin eastward into Manchuria; there are large industrial towns along the Peking-Hankow railway and in Shantung. Among the coalfields on the northern edge of the area, the Kailan Basin has the highest output in China. This coal is distributed by railway, and through the ice-free port of Chinwangtao. Of the smaller fields along the Peking-Hankow railway, those of Lincheng and Fengfeng supply coking coal to iron and steel works northward as far as Anshan in Manchuria.

Tientsin (population 3,220,000), 40 miles from the sea on the shallow Hei Ho which freezes in winter, can be reached only by small vessels. Its outport, Tangpu, accommodates ships of 10,000 tons and can be kept open by ice-breakers during the two months of the winter freeze. Opened by the Chinese to international trade about 150 years ago, Tientsin flourished despite the navigation difficulties. The hinterland includes Inner Mongolia and reaches far into Sinkiang. Among the industries of Tientsin, food-processing predominates, but steel, chemicals and engineering works are also important.

Tsingtao (population 850,000 in 1948) has the advantage of a deep-water harbour on the rocky Shantung coast and a good route through the Kiaolai corridor to the North China plain, where sedimentation and shifting river-mouths make port construction and maintenance difficult and expensive. The place was of minor importance until Germany held it as a commercial base from 1898, when German development made it one of the best ports in China. Improvements continued after 1914 under the Japanese, who returned the city to China in 1924. There are heavy-engineering, steel and light industries; coal comes from several fields on the north and south-west borders of the Shantung highland. The Shantung peninsula abounds in good harbour sites, but only Tsingtao has achieved major importance. Wei Hai, formerly Weihaiwei, was a British naval base from 1898 until 1930.

Peking (population 5,420,000), an ideal site from which to control the Mongolian borders, was the Manchu capital. Until the Communists selected it as their capital in 1949 it was an administrative, cultural and communications centre. It is now also a modern industrial city. Its iron and steel plant, 12 miles away on the Yungting river at Shihkingshan, supplies pig iron to steel works at Tangshan and Tientsin.

Kaifeng, capital of China for 200 years in the tenth and eleventh centuries, is of little significance today in comparison with industrial Chengchow about 40 miles to the west, where the Peking-Hankow railway crosses the Lunghai railway that runs west to Paoki and Lanchow. Loyang, another former capital, owes its modern growth to the establishment there in 1955 of the first modern tractor plant in China. Tsinan is a railway focus with some industry, hampered by the poor conditions for river shipping on the Hwang Ho.

(4) THE YANGTZE RICE/WHEAT AREA

In this region most of the fields carry two crops a year: rice or cotton in summer, and wheat, maize or barley in winter. Sometimes a catch crop of vegetables is also obtained, making three crops a year all from the same ground. Cotton is particularly prominent on calcareous alluvial soils near the coast which are too alkaline for rice. About two thirds of the land is irrigated, mainly from small streams, canals and ponds. On the hillsides tea and mulberry are characteristic crops. The average rainfall of 60 in. in the south, decreasing to about 30 in. in the north, is not only higher than in north China but also better distributed throughout the year. Frost occurs every year, but sustained freezing for more than two or three days is rare. The northern boundary corresponds broadly with the northern limit of leached soils: rice is grown beyond it but does less well than to the south.

The physical components of the region are the southern edge of the North China plain, all the Yangtze plain and some bordering highlands in the west. The Yangtze plain, in which population densities exceed 2,000 people to the square mile over large areas, widens in places to 200 miles; elsewhere, the hills come close to the river. Like the Hwang Ho, the Yangtze is subject to annual floods, but it is less of a menace to life and property because several lakes along its course help to regulate the flow. The largest of these lakes, Tungting and Poyang, are about 100 miles across at high water.

Shanghai (population 6,204,000), built on a mudflat on the left bank of the Whangpoo river about 15 miles from the Yangtze and 50 miles from the open sea, was opened to foreign traders in 1842. An international settlement with modern buildings and wide streets grew beside the old town. Though constant and expensive dredging is necessary to keep the river deep enough for ocean shipping, over half the Chinese imports entered this way during the years before 1939. This trade has suffered with the reduction of imports from western countries under the Communist régime, but Shanghai remains an important industrial centre, with textile factories, heavy engineering, shipbuilding and steel works. An oil refinery draws some of its crude oil from the Yumen fields.

Nanking (population 1,419,000 in 1960), 150 miles up the Yangtze, was the capital of China under the Nationalist régime from 1928 until 1937. Five hundred miles from the sea and accessible to ocean shipping, the three towns of Hankow, Hanyang and Wuchang constitute the conurbation of *Wuhan* with an aggregate population of 2,146,000 (1957) and a wide range of heavy and light industries, including chemicals, textiles, cement and the new iron and steel works. A major achievement of the Communist régime was the successful bridging of the Yangtze at Wuhan in 1957. The bridge, which carries a double-track railway and a six-lane motorway, provides 56 ft. clearance above the high-water level so that ships as large as 10,000 tons can get by.

Coal is mined at several of the scattered fields indicated on Figure 31, but principally at Pingsiang, where high-grade coking coal is worked for the iron and steel industry of Wuhan, and around the new city of Hwainan, built after 1949 on the south bank of the Hwei Ho. The Hwainan field is expected to produce seven million tons a year. Some of the coal is taken south by rail and shipped from the Yangtze port of Yukikow opposite Wuhu.

(5) THE SZECHWAN RICE AREA

This region is characterized by a greater range of crops than any other part of China. Rice is the main one, but maize, sugar cane, cotton, tea, tobacco, wheat, millet and kaoling are all prominent—the wheat and millet especially so on the higher ground. The annual crops are managed so that practically every field gives two or sometimes three harvests a year. Tung oil, from both wild and cultivated trees, is an export crop. The boundaries correspond broadly with those of the physiographic division known as the Szechwan Basin. This basin is drained by the Yangtze and its several left-bank tributaries, among which are the Min Ho and the Kialing. Most of the country is hilly, rising from 1,000 ft. or so in the central parts to mountains of 10,000 ft. in the north and west and 8,000 ft. in the east. Hillsides in the soft red sandstone, which is the surface rock over large tracts, are generally steep. Away from the flood plains the only flat or gently undulating land is on the rounded hill-tops.

The sandstone weathers to red and purple soils, hence the name Red Basin of Szechwan that is sometimes used. The largest expanse of alluvial lowland is the Chengtu plain, some 70 miles north to south and 50 miles across, on the Min Ho.

The basin is more effectively shielded by mountains from the influence of cold continental air in winter than is the lower Yangtze valley, consequently it has a milder climate. Frost is rare at levels below 3,000 ft. Most of the rain comes in summer, but there is some in every month of the year. High altitude and very steep slopes preclude agriculture from all but a fifth of the total area, but that fifth is probably more densely settled and intensively cultivated than any other part of China. The irrigation canals which support the tremendous crop yields of the Chengtu plain have been working for over 2,000 years. Throughout the region the pressure of population on land is such that paddy is grown on terraced and irrigated slopes steeper than 45 degrees: the drop from one field to the next is usually vertical and greater than the width of the fields. Under these conditions much of the farmers' time is taken by repair work.

The Szechwan Basin (Figure 37), with about 10 per cent of the total Chinese population, is isolated by mountainous country from the other populous parts of China. The 500-mile stretch of the Yangtze used by 500-ton steamers within the region is separated from the easily navigable lower reaches by several gorges, where the current is fast and the difference between low water and flood level might be 200 ft. Dangerous and difficult as the gorges are, river shipping manages to get through to Chungking. Small, powerful steamers, cleverly handled, can just hold their own. Sailing junks are towed upstream by teams of men who haul long ropes of split bamboo along the riverside tracks. The larger boats might require 200 men for this task, taking four to eight weeks to make the 100-mile passage above Ichang. Professor Cressey records that a river steamer was once stranded on a rock 120 ft. above low water.

Chungking (population 2,121,000 in 1957) rose to importance as the capital of the Nationalist Chinese during World War II. In that period an iron and steel industry was founded, the silk and cotton industry expanded and coalfields both near the city and south of the Yangtze were opened up. There is a substantial production of salt in the form of brine from deep wells in the tributary valleys leading south to the Yangtze. Entirely dependent until 1956 on the Yangtze route for communication with eastern and northern China, Szechwan now has two useful railway

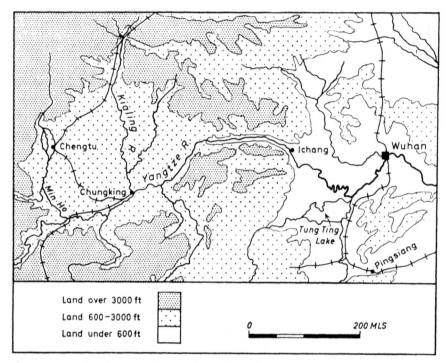

Figure 37. *The Szechwan Basin and the middle Yangtze*

links—one northward by the city of Chengtu to Paoki, the other south-ward to Kunming.

The administrative unit, the province of Szechwan, roughly coin-cided with the agricultural region until 1955. In that year the province was enlarged to embrace the adjacent territory of Sikang on the edge of the Tibetan plateau.

(6) THE RICE/TEA AREA

Here the farming scene is dominated by rice on all the lower ground and by tea on the hillsides. After rice the principal food crop is the sweet potato. Other crops are citrus fruits, sugar cane and tobacco. Wheat, maize and millet are grown in the higher districts. Apart from the fringe of the Yangtze plain and the basin of Poyang Lake in the north, there are no large alluvial areas. It is a mountainous region. Not more than 20 per cent of the total area is farmed; the other 80 per cent is too hilly. In the centre and east, ranges roughly parallel to the coast rise to 5,000 ft. Parts of the Kweichow plateau at 4,000 ft. lie within the western border. Rainfall is plentiful and well distributed throughout the year. On the cultivable ground densities of population are among the highest in China. Pressure is so great that many make their living from the sea; fishing and coastal trading are both important. For centuries there has been a tradition of emigration to Malaya and other countries of South-east Asia.

The coast is rocky, with many islands, and abounds in good natural harbours. Yet none of these harbours has gained more than local importance, because their natural hinterlands are restricted by the steep and parallel ranges near the coast. The largest port, Foochow, exports tea. At the close of the nineteenth century the Chinese contributed nearly half the world's tea exports. Today the chief suppliers of the much larger world consumption are the efficient plantations of Ceylon and India; most of the Chinese tea is consumed in China.

Internal communications in the rice/tea area are poor. Inland the hills are well timbered. Lumbering is an important industry. Fir and pine yield structural timber. Camphor and tung oil are other forest products.

(7) THE DOUBLE-CROPPING RICE AREA

This region is partly within the tropics. In the south three crops of rice are taken from the same ground each year. In the north two rice crops are usual, followed by a third crop of vegetables. Other crops are the same as in the rice/tea area, with the additions of coconuts, bananas and rubber. The country is as mountainous and forested as the rice/tea area, and only about 13 per cent is cultivated. Movement is easier because the Sikiang and its tributaries, which converge on the Canton delta, provide navigable routes westward to the Vietnam border and northward for nearly 200 miles; 2,000-ton river steamers can reach Wuchow.

According to Theodore Shabad, the 3,000 square miles of the Canton delta—the largest alluvial tract in the region—support about 10 million people, including the urban population. Among the industries of Canton and its outport of Whampoa (combined population 1,840,000 in 1957) are engineering works and cotton, silk and jute mills. Until recently the port could not handle ships larger than 5,000 tons; as a centre of foreign trade it is dwarfed by Hong Kong less than 100 miles to the south-east. Of the several small ports between Canton and Shanghai, only Amoy is served by a railway. The main purpose of this line, which was completed only in 1957, was to supply military bases on the coast opposite the Nationalist strongholds of Taiwan and the smaller islands of Quemoy and Matsu.

The coastal areas towards the Vietnam border are rich in manganese, while the island of Hainan has impressive reserves of iron ore. Canton has railway connection with Hong Kong and northward to Wuhan.

(8) THE SOUTH-WEST RICE AREA

This area differs from the other rice-growing regions of south China in that rice can be grown only as a summer crop. Winter crops are wheat, millet and maize. Rice cannot be raised there satisfactorily as a winter crop, despite the southern location, because winter temperatures are too low. Most of the area is plateau at a general level of 6,000 ft. in Yunnan and 4,000 ft. in western Kweichow; the floors of some of the river valleys fall to 2,000 ft. but only a very small proportion of the area approaches that level. There are extensive tracts of karst country.

In the western part the headwaters of the Red river (of Vietnam), the Mekong and the Salween have cut valleys too gorge-like in character to provide good north-south routes and so deep as to make east-west roads and railways expensive to build. Probably less than 10 per cent of the total area is cultivable.

Remote from the rest of China Proper, this is the portion most recently colonized by the Chinese. Only half the population is Chinese in speech. The remainder are indigenous people in process of assimilation into Chinese culture.

The railway to Hanoi from Kunming, the capital of Yunnan, and the routes for pack-animals into Tibet have never been of great economic significance. The political and strategic importance of the area arises from its location on the Burma border, and astride the only practicable (though immensely difficult) land route into China from the western world during World War II. The Burma Road—a track with Alpine gradients but fit for robust motor trucks—between Chungking and the Burmese railhead near Lashio was opened in the 1930's. Though plans to provide a railway link with Lashio were abandoned in 1940, the war resulted in industrial development and population growth at Kunming.

The natural resources include valuable deposits of tin, copper, mercury and scattered coal seams. Tin was smelted and exported by rail through Hanoi until 1940, and the mercury of Kweichow reached world markets. Mineral production and timber-felling in the extensive forested areas are expected to increase as rail and road communications improve. It is doubtful whether trade with Burma will justify the maintenance of the road to Lashio at a standard fit for motor traffic.

TAIWAN (FORMOSA)

From the agricultural viewpoint, Taiwan is part of the double-cropping rice area of southern China. A distance of 250 miles from north to south, about 100 miles wide and lying 100 miles off the mainland, the island has a mountainous eastern section rising to 13,000 ft. and a wide coastal lowland with extensive alluvial spreads on the west. On the east the seaward slopes are steep and there is practically no coastal lowland. Rainfall in the highlands reaches 200 in. In the western lowland, sheltered from the full effects of the summer monsoon by the mountains, irrigation is generally practised because the annual rainfall might be as low as 45 in. and the evaporation rate is high.

Taiwan first became a province of China in 1683 under the Manchu régime. The island prospered under the progressive agricultural policy of the Japanese, who acquired it after the Sino-Japanese war of 1895 and held it until it was returned to China in 1945. The Japanese encouraged careful manuring with natural and artificial fertilizers; surplus rice and sugar were exported, mainly to Japan. But Japanese farmers were unable to establish themselves in Taiwan because they could not compete economically with the hard-working Chinese settlers. The Japanese settlers, who were entirely urban dwellers engaged in industry and administration, were repatriated to Japan under the terms of the peace arrangements of 1945.

During the last 20 years of the Japanese period there was considerable industrial development, especially at Takao in the south and at the capital, Taipei. Large coal reserves of rather poor quality were worked in the north; production today is at the rate of four million tons a year. Copper and gold are mined. Of the hydro-electric potential of 3,300,000 kW, about 10 per cent is developed.

An impression of the economic development of Taiwan in comparison with the mainland can be gained from the following figures: Taiwan, with an area of only 36,000 sq. km., has 1,000 km. of railway; the whole of mainland China, with 240 times that area, has only 30,000 km. of railway.

In 1949 the defeated forces of the Nationalist Chinese withdrew to Taiwan from the mainland. During the period 1949 to 1960 their propaganda envisaged an invasion to recapture the mainland, where, it was hoped, the peasants would rise against the Communist authorities. The population is about eight million, and rural densities on the best land are as high as anywhere on the mainland.

OUTER CHINA

Throughout history the inner Asian boundaries of Outer China have shifted with the relative strengths of the Chinese central governments in their administration of distant possessions on one side and the determination of local non-Chinese populations (often assisted by foreign powers) to assert their independence on the other. Chinese authority reached its maximum extent during the Manchu period. At that time it embraced all Manchuria, all the country north of the Great Wall as far as the borders of Siberia and all the land westward to the Pamirs, including Tibet. In this enormous area five physical constituents are especially significant in the political evolution (Figure 38):

1—A belt of desert and arid scrub extending from the western end of the Tarim and Dzungarian basins through the Gobi Desert to the borders of Manchuria.

2—North of the desert, a wide belt of grassland, with some forest near the border of Siberia.

3—South of the desert, another belt of grassland, 150 miles wide in the east where it lies to the north of the Great Wall, but becoming very narrow in the west where it is confined between the Taklamakan Desert and the high, barren slopes of the Nan Shan and Altyn Tagh.

4—The inhospitable Tibetan plateau, much of it at 15,000 ft., and practically all above 10,000 ft., apart from a few valley bottoms in the east and south. The maximum width of the plateau between the bordering ranges of the Himalayas in the south and the Altyn Tagh in the north is 700 miles.

5—The oases. Both the Tarim and Dzungarian basins are areas of inland drainage. Most of the rivers fed by melting snow on the adjacent mountains have too little water for them to persist far into the desert, but they are all of great importance because they bring moisture to the desert margins. Where each stream dwindles into the desert there is an oasis. Thus there is a ring of oases around the Dzungarian Basin and another around the Tarim Basin. The line of oases along the southern edge of the Tarim Basin continues eastward along the foot of the Nan Shan into Kansu.

The Chinese did not have the spare manpower to garrison and administer so much country. They gave most of their attention to areas of special strategic interest and economic value.

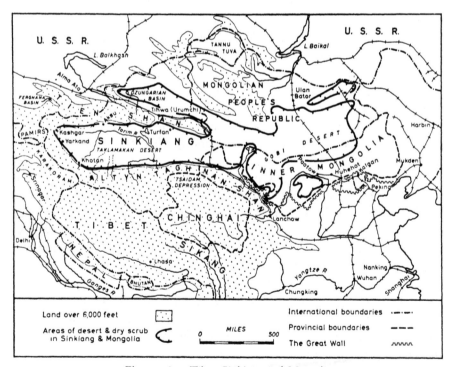

Figure 38.   *Tibet, Sinkiang and Mongolia*

Mongolia, the 900 miles of country extending from the Great Wall northward to the border of Siberia, was divided into inner and outer zones relative to China Proper. In Inner Mongolia, which included the grassland adjacent to the Wall as well as the southern edge of the desert, government was firmly exercised by Chinese officials, and Chinese farmers were encouraged to settle there wherever there was irrigable land among the pastures of the Mongol herdsmen. Outer Mongolia, which included most of the Gobi Desert and all the northern grassland and forest, was too remote from China Proper to constitute an invasion threat, and too poor in resources to justify an expensive administration. The local chiefs of the pastoral tribes were left to govern themselves, provided they acknowledged the suzerainty of the Manchu rulers.

This political division into inner and outer territories was also applied to the Tibetan plateau. Inner Tibet, the part nearest China Proper, was governed by the Chinese. Outer Tibet, which is the area generally called Tibet by western people, was left to look after itself. A political agent and a small garrison were stationed at Lhasa as a sign of Chinese authority, but the Tibetans generally took little notice of them.

A division into inner and outer zones was considered unsuitable for the north-western sector between Tibet and Mongolia. The lines of oases around the desert basins determined the caravan routes to the west; many of these oases were large and prosperous, and the central Asian routes acquired a prestige value out of proportion to the trade they carried. The Chinese established direct civil and military control in this area and gave it the name of Sinkiang (or New Dominion) when provincial status was granted in 1878.

Throughout these areas Chinese domination was resented. Armed rebellion was frequent. After a series of revolts in which local forces received Russian help, Outer Mongolia became in 1921 the independent Mongolian People's Republic. In the early decades of the present century Chinese control in Tibet was so weak that Tibetan leaders in Lhasa made their own political and trade agreements with India. Thus Tibet became widely recognized as an independent country closely associated with India. In recent years Indian influence in Tibet has weakened. Chinese armies captured Lhasa in 1951 and now control the whole country right up to the Indian border. The Mongolian People's Republic, backed by the power of neighbouring Russia, is likely to remain free of Chinese control.

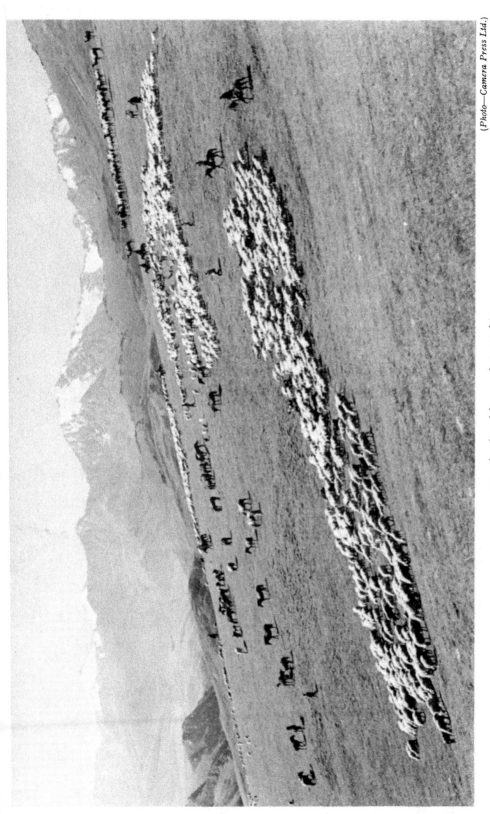

13. Grassland and livestock in Sinkiang

14. Malaya: aerial view of a tin dredge

15. Malaya: three dredges working tin-bearing alluvium near Kuala Lumpur

Unlike the other great lowlands of Monsoon Asia, the plain of Manchuria (Figure 39) is the product of erosion, not of recent deposition. In the riverine tracts there are alluvial spreads, but most of the plain is gently rolling country with hills rising to several hundred feet. The bordering highlands on the north, east and west maintain an average level of 3,000 ft. with a few peaks approaching 10,000 ft. Average rainfall is about 40 in. in the eastern highlands, 30 in. in the eastern portion of the plain, falling to 15 in. in the west and north. Average daily temperatures below zero persist for three months in the south and six months in the north; July mean temperatures range from 76°F. in the south to 70°F. in the north.

This harsh environment held little attraction for the early Chinese. Their tradition of intensive agriculture associated with high rural population densities and great social cohesion demanded a milder climate. In most of Manchuria comparatively few people could live off each square mile. This meant dispersal of population at much lower densities than those prevailing in the North China plain, less social cohesion and greater difficulties in binding scattered farmers into easily administered groups. Settlement in Manchuria was therefore not greatly encouraged by the Chinese rulers. By the Ming dynasty (1368–1644), population pressure in the North China plain had resulted in Chinese emigration to the Liaoning plains, but most of Manchuria was left to nomadic tribes. A big influx of Chinese farmers began 60 years ago. All the best land is now occupied, but there are still some opportunities for extension of cultivation in the north; farms are larger than in China Proper, and modern agricultural machinery is used. There is surplus grain for export. Main crops are soya beans and kaoling, with smaller acreages of millet and spring wheat. In the moister Liaoning plains cotton, corn and even some rice are grown. The soya bean is a valuable commercial crop, yielding human food, cattle cake and oil, which is used in chemical industries.

A good railway network, the best in China, facilitated rapid settlement, industrial growth and the export of agricultural and industrial products. The first railway concession was granted by the Chinese in 1896 to the Russians. Russian interest was mainly strategic, for Manchuria formed a huge salient of Chinese territory astride the best route from the Lake Baikal area to Vladivostok, but there was also provision for the

Russians to work the mineral resources. The city of Harbin was built where the Russian line crossed the Sungari river. This line continued southward into the Kwantung peninsula. Dairen grew as a commercial port, and Port Arthur as a Russian naval base. After the Russo-Japanese

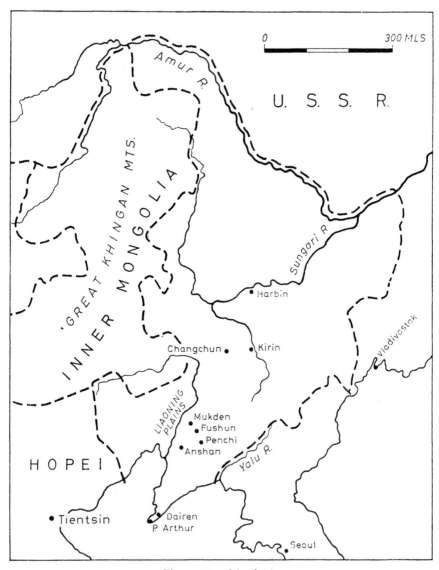

Figure 39. *Manchuria*

war of 1904-5, Japan became the dominant foreign influence and many more railways were built. The Peking-Mukden line was a British venture.

Much of the industrial growth took place during the period 1931 to 1945, when Manchuria was occupied by the Japanese and known as Manchukuo. Japanese farmers were unable to settle in the country because they could not compete with the Chinese and Korean settlers, who tolerated very low living standards, but Japanese capital investment brought spectacular progress in mining, manufacturing, the building of hydro-electric works on the Sungari and the general development of the area as a market for Japanese manufactured goods. In 1945 there were 750,000 Japanese administrators and industrial workers in Manchuria: all were repatriated under the terms of the peace settlement.

Although other parts of the country are richer in natural resources, Manchuria is today the leading industrial zone in China. Mukden (population 2,411,000), Dairen/Port Arthur (population 1,054,000 in 1950), the great steel centre of Anshan, Harbin and Changchun are major industrial towns. Fushun and Penchi are two of the many smaller industrial centres. Kirin has timber-based industries supplied by the forests of the eastern highlands. The combined activities of all these places cover a remarkable range of industry embracing iron and steel, engineering, motor cars and trucks, chemicals, textiles, oil-refining and shipbuilding.

(2) INNER MONGOLIA

During the Manchu period Inner Mongolia extended from the Kansu border eastward over the southern fringe of the Mongolian plateau at a general level of 3,000 ft. to the heights which overlook the plains of Manchuria. Under the Communist régime it became a self-governing unit, the Inner Mongolian Autonomous Ch'u. In 1949 the area was extended eastward to include some arid country and the Great Khingan mountains, both formerly part of western Manchuria.

In the Khingan mountains rainfall is sufficient for forest growth. The remainder of Inner Mongolia has too low and unreliable a rainfall for agriculture to flourish without irrigation, and on the northern borders the poor grassland fades into stony scrub and desert. There are desert tracts also in the Ordos plateau, south of the great northward bend of the Hwang Ho. The indigenous Mongol population, perhaps a million in all, have largely abandoned their former livelihood of nomadic pastoralism in favour of settled agriculture in a few irrigable districts. Now probably no more than 300,000 move around the pastures with herds of sheep and goats together with some cattle, horses and camels. Chinese farmers number about 80 per cent of the total population of six million. Together with the Mongol farmers they cultivate the irrigable land along the Hwang Ho and the tributary valley leading to the capital, Huhehot. Crops are those of the spring-wheat area (see page 175). There is some scope for extension of irrigation along the Hwang Ho. Pastoralism is the only practicable rural economy in the remainder of the area, except in the Khingan highlands, where lumbering is a growing industry in parts accessible to the railway linking Manchuria with Siberia.

Political and economic cohesion in Inner Mongolia is hindered by the absence of good communications linking outlying areas with the capital. In several areas isolated from one another but with rail links southward to China Proper, industrial progress has been made: outstanding is the mighty iron and steel project at Paotow. The completion in 1956 of the railway across the Gobi Desert to the Trans-Siberian line made Tsining the junction where goods and rolling stock are transferred from the Chinese standard gauge to the Russian broad gauge. Iron ore and other mineral deposits in the desert will probably be opened up along this Trans-Mongolian railway.

(3) THE MONGOLIAN PEOPLE'S REPUBLIC

Some account of this independent state in a chapter on China is appropriate on grounds of the former association of Outer Mongolia with the Manchu empire. The grasslands and forests of the north, the only habitable parts of Outer Mongolia, are separated from China by several hundred miles of desert but have easy routes to Siberia. Strong economic ties with the Soviet Union are therefore to be expected, and it is understandable that the Chinese should have failed to maintain their hold on the country. Arable farming is precluded by the severe climate. The only practicable rural occupation is stock-raising. In recent decades haymaking and the provision of more wells have enabled the Mongol people to abandon nomadism in favour of a more settled existence. Ulan Bator, the capital, has leather industries and railway communication with both

China and the U.S.S.R. In the north-east several branch lines from the Trans-Siberian railway penetrate to mineralized areas deep in Mongol territory but no details of mining are available.

(4) INNER TIBET

Before 1955 this area consisted of two provinces: a northern one, Ching-hai; and a southern one, Sikang. In that year Sikang was abolished; the eastern portion was merged with Szechwan and the remainder was included in the outer territory of Tibet. Despite this political rearrangement, it is geographically sound to retain Sikang as a regional name for the south-eastern edge of the Tibetan plateau. There the plateau at a general level of 12,000 ft. is dissected by the headwaters of the Yangtze, Mekong and Salween. Chinese have penetrated to the cultivable valleys, some of which fall to 5,000 ft.; on the higher levels people of Tibetan and other tribes live by primitive farming and stock-raising. The overall population density is very low. In Chinghai the plateau surface is less dissected; agricultural opportunities are even poorer and there are fewer people. In the Tsaidam depression there are encouraging signs of oil and other minerals to which the Chinese plan to build a railway from Lanchow.

(5) TIBET

The Chang Tang, the plateau at a general level of 15,000 ft. occupying all central and northern Tibet and amounting to three quarters of the total area, is uninhabited, and practically devoid of vegetation. The mean annual temperature is about 25°F., and there is severe night frost even in June. Such a large area might be expected to contain useful minerals, and the Chinese began prospecting around 1955. To the south is the long and shallow depression at about 12,000 ft. drained by the headwaters of the Brahmaputra (called the Tsangpo in Tibet) and the Indus. Southward again to the Indian border are the Himalayan ranges with peaks over 20,000 ft., great plateau tracts at 12,000 to 15,000 ft. and many gorges with floors cut down to 8,000 ft. These gorges drain southward through the Himalayas to the Indian plains. Almost the entire population of around 1,270,000 live in this lower quarter of the country to the south of the Chang Tang where the climate is milder. Winters are too cold for autumn-sown crops, but summers are warm, with a modest rainfall from the summer monsoon. Barley is the chief grain, raised in small irrigated fields. Favourite cultivation sites are alluvial cones along valley sides. Herds of sheep, goats and yaks, with some horses and camels, are important in the economy. The only export of any economic account is wool.

Until the Chinese invasion of 1951 Lamaist Buddhism was the most powerful social force, and the highest ambition of Tibetan families was for their sons to join one of the several great monasteries. The monasteries owned much agricultural land, their income from that source being augmented by gifts from the entire population. The Dalai Lama, the head of the Buddhist priesthood, was also the temporal ruler. He and his government occupied the Potala, the great monastery which dominates Lhasa. Great power was exercised by local chiefs and landlords.

Although the Chinese never abandoned their claim to Tibet, the country was virtually independent from the closing years of the Manchu period until the departure of the British from India in 1947. Both the Manchus and the Nationalists failed to maintain order in Inner Tibet. Traders moving to eastern markets through Chinghai and Sikang were likely to suffer at the hands of bandits; it was quicker and safer for the Tibetans to take their wool by pack-mule across two 15,000 ft. passes on a track from Lhasa to Kalimpong in the Bengal foothills, whence there is a railway to Calcutta. That journey took only 14 days.

This diversion of Tibetan trade towards India was the result of the border policy of the British Indian government. While there was no great military threat to India from the direction of Tibet, two considerations gave grounds for anxiety about the safety of the Himalayan frontier. First, unrest among the Tibetan people might have necessitated British garrisons along the whole of the 1,600-mile border, and the cost would have been insupportable. Second, Chinese settlement in the valleys of Sikang was taken as a sign of mounting population pressure which might ultimately have endangered the relatively empty lands of Assam. After the failure of diplomatic efforts to gain Chinese and Tibetan agreement to restrict Chinese settlement and allow British political and trade agents to be stationed in Tibet, a small military force was sent from Bengal in 1903 under Colonel (later Sir Francis) Younghusband. Sharp Tibetan resistance was overcome. When the force reached Lhasa in 1904 the

Dalai Lama had gone, but his ministers agreed to the British demands. Henceforth, the British were to be the dominant foreign influence in Tibet, British political officers and small groups of Indian troops were to be stationed along the main trade route between Lhasa and Kalimpong, and future Chinese settlement was to be confined to Inner Tibet. Under these conditions, ratified by representatives of the three powers at the Simla Conference in 1914, the trade with India prospered until 1947. The new independent government at Delhi, fully occupied with the internal problems following the partition of India, was unfortunate in coming to power at a time of renewed Chinese pressure. After capturing Lhasa in 1951 the Chinese extended their military roads to several points on the Indian border. The Indian troops were withdrawn. All the trade was diverted along the new roads to Chinese railheads. Only Tibetan refugees from Chinese oppression came over the passes to India.

There is no sign that the Delhi government will re-establish Indian influence in Tibet. The Chinese now describe the country as the Tibet Autonomous Ch'u. How much authority remains in the hands of the monasteries and the priests is not clear. The Chinese intention is ultimate assimilation into the Communist way of life.

(6) SINKIANG

The Tarim and Dzungarian basins, arid and bordered by high ranges, take up most of the area. Of the total population of just under six million, 75 per cent are Uighers, Muslim farmers living in the oases; 8 per cent are Chinese farmers and traders, and the remaining 17 per cent are nomadic pastoralists. Among the Chinese a small Muslim minority, the Tungans, have traditionally played a leading part in revolts against the Chinese government.

The largest oases are Kashgar, Yarkand, Aksu, Khotan, Urumchi (Tihwa) and Turfan. All these lie between 2,500 ft.–4,500 ft. except Turfan where the ground sinks in a narrow depression to 427 ft. below sea level. Grain crops, cotton and a great variety of fruits are raised. The old caravan routes between the oases are now motor roads suitable for heavy trucks; communication with adjacent parts of Russia is easier than with eastern China. The Chinese and Russians are co-operating in the development of mineral resources along the flanks of the Tien Shan; oil, coal and non-ferrous metals have been found so far. The area is destined to become a prominent industrial zone when the railway from Kansu through the capital, Urumchi, to the Soviet network in Russian Turkestan is completed.

CHINA: THE COMMUNIST EFFORT

Under the Communist régime tremendous progress is being made in agriculture and industry. Many social functions of the family have been taken over by the People's Communes: in rural areas a commune is intended to embrace about 2,000 families. Details of farming routine and the allocation of specific tasks in the fields are decided by a people's committee at the commune headquarters in the main village of a district. The small plots of the days of individual ownership have been merged to form fields large enough for mechanized agriculture when more machines become available. By 1960 the annual income of the peasant was three times the 1939 figure. Food production is improving, partly as a result of better seed and manures and more irrigation; and also because communal feeding and care of infants in the communes frees additional labour for work in the fields. The commune system also operates in many towns, particularly where there are new factories. It remains to be seen whether the old traditions of family loyalty, classed as reactionary by the Communists, will die with the passing of the present older age groups. It is possible that the present younger generation, now at least in appearance strongly in support of the government, might come to regard Communist living as too high a price for material progress. On the other hand, Communism might provide the best conditions for the rapid economic improvement the Chinese need. A fully industrialized China, with a rapidly increasing population of over 600 million, under a totalitarian régime would bring drastic changes to the world balance of power.

HONG KONG

The hilly island of Hong Kong was granted to Britain by China in 1841. The mainland peninsula of Kowloon was similarly granted in 1860. The island and peninsula together form the British colony of Hong Kong with a magnificent natural harbour and a total area of only 35 square miles (Figure 40). In addition there are the New Territories, about 350 square miles of adjacent mainland and numerous islands, leased in 1898 to Britain for 99 years. Under stable government Hong Kong became a great commercial centre and attracted Chinese traders from the mainland.

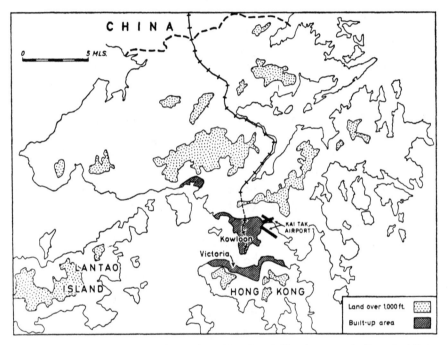

Figure 40. *Hong Kong* Hong Kong island and Kowloon are indicated. All the other islands, together with the mainland beyond Kowloon and as far as the China border, constitute the New Territories. Note the long runway of Kai Tak airport extending into Kowloon Bay

The population grew from around 2,500 in 1845 to 1,640,000 in 1941. On Christmas Day 1941 the colony surrendered to the Japanese, but many Chinese managed to escape. In 1945 probably no more than 500,000 people remained there. Economic recovery after the war was rapid. By 1961 the population was 3,178,000, all but 20,000 of whom were Chinese. A high proportion were destitute refugees from the mainland; living in rough huts crowded together, they became a menace to public health. Government housing and settlement schemes were inaugurated to cope with this problem: water supply is one major difficulty; in 1957 it was restricted to two and a half hours daily during the dry season.

Since the Communist victory in 1949 trade with mainland China has declined and the prosperity of Hong Kong has become increasingly dependent on manufacturing industry. Of the requirements for industrial progress, the natural resources provide almost nothing. There is plenty of labour. Capital is invested by Chinese and foreigners because the stable government of the British is still the best guarantee of commercial security in Monsoon Asia. Only 12 per cent of the land is cultivable, and that mainly in the north of the New Territories. 80 per cent of the population is urban. 90 per cent of the food is imported, together with all the coal, oil, cotton, steel and timber. The range of industry is astonishing. Of the 150,000 industrial workers, 30 per cent are in textiles. Shipbuilding, light engineering and the making of consumer goods as diverse as electric torches, pressure cookers, toys and plastic goods occupy most of the remainder.

Victoria, the main settlement on Hong Kong island, is a city on the northern slopes of the Peak, which rises to 1,800 ft. On the mainland across the harbour is the town of Kowloon. Flat land is so scarce that large coastal reclamation schemes are undertaken, of which the most spectacular is the enlargement of the Royal Air Force station of Kai Tak to take the world's biggest commercial planes. Surrounding hills precluded the extension of the old runways, so a new runway two miles long was built out into Kowloon Bay. Over 100,000 Chinese live on boats moored in the harbour. Sea-fishing is an important contribution to the food supply.

MACAO

This Portuguese colony of only six square miles lies 35 miles west of Hong Kong near the mouth of the Sikiang. Chinese merchants conduct a small but varied local transit trade. The harbour depth is only 15 ft.

14

Japan and Korea

THE PHYSICAL BACKGROUND TO AGRICULTURE AND
SETTLEMENT IN JAPAN

The four main islands of Japan are mountainous; their structure is complicated, but in the south two parallel mountain alignments are clear, with the depression between them occupied by the shallow Inland Sea. There are active volcanoes in Kyushu, central Honshu and Hokkaido, and a great deal of the high ground is built of volcanic material unfortunately acid in composition and therefore associated with poor soils. Two thirds of the entire area is too steep for cultivation. Much of the remaining third has slopes steep enough to be agriculturally useful only under careful terracing. The only large alluvial lowland is the Kanto plain of about 5,000 square miles extending for 80 miles northward from Tokyo. The country abounds in smaller alluvial lowlands, mostly coastal or on the lower reaches of rivers and separated from one another by mountain ranges.

The latitudinal extent of over 1,000 miles, the complicated relief and the variety in exposure to the main sources of weather combine to give a striking range of climatic conditions with sharp changes within short distances. Professor L. D. Stamp recognizes four broad climatic divisions (Figure 41):

1—*Southern Japan*, a sub-tropical area including Kyushu, Shikoku and the southern fringe of Honshu. The average annual rainfall reaches 100 in. on slopes exposed to the summer monsoon, and exceeds 40 in. everywhere except in the rain-shadow belt along the shores of the Inland Sea. The frost-free season on the lower ground is about 240 days.

2—*Eastern Japan* includes the eastern part of Honshu to the north of Tokyo, and a small part of southern Hokkaido. Summer temperatures are as high as in southern Japan, but the winters are cold and dry. January mean temperatures are below freezing in the northern half, and not much above in the south. The frost-free season is about 240 days in the south and 150 days in the north.

3—*Western Japan* includes all Honshu to the west of the main crestline, together with western Hokkaido. The climatic feature common to the whole is substantial winter rainfall brought by the winter monsoon. Almost everywhere the annual rainfall exceeds 60 in., and is well distributed throughout the year with a prominent winter maximum and a secondary summer maximum. The frost-free season increases from an average of 141 days in the north to 206 days in the south.

4—*Northern Japan* covers the northern portion of Hokkaido. Summers are cool, the winters cold and the frost-free season about 140 days. Annual rainfall is generally below 40 in.

Throughout Japan, summer conditions permit the cultivation of rice. Even in Hokkaido rice is the most important single crop, though yields per acre there are lower than in the south. Winters are warm enough to allow two rice crops a year only in the south (Figure 41), but winter crops such as wheat, barley and vegetables are planted after the rice harvest in Kyushu, Shikoku and the southern part of Honshu. In the southern part of Japan, where the climate is most favourable for intensive farming, much of the ground is high and agriculturally useless. Over a quarter of the lowland is in Hokkaido where winter crops cannot be grown.

JAPAN: HISTORICAL GEOGRAPHY

Japanese culture appears to have developed first in the south-east and to have spread northward through Honshu. The beginnings in the south owed much to contacts with China during the Han period, but the nascent Japanese civilization was not a mere copy of the Chinese. The Japanese rather adapted Chinese ideas to fit their own needs and environment. Intensive agriculture required little alteration; Chinese social and

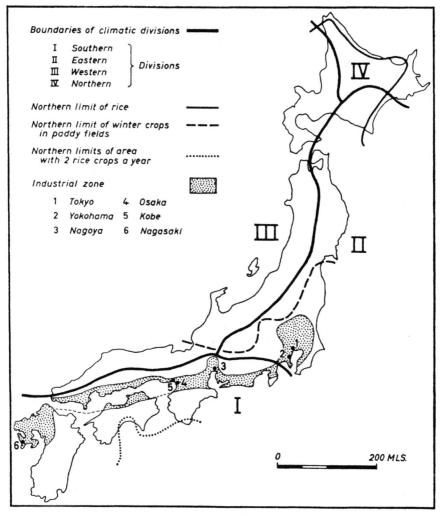

Figure 41. *Japan*

political organization was another matter, and in that sphere the Japanese had to be selective. Wholesale imitation would have meant concentrating political power in a single ruling authority. Such a prospect had no appeal to the Japanese, divided as they were into clans each reluctant to see its neighbours gain too much power. Indeed, it is unlikely that a single centralized state would have worked in those early days, even had the Japanese feudal clans wanted it, for there was no physical counterpart of

205

the rich and extensive North China plain in which an enormous population could be governed easily from some central point. The agriculturally desirable parts of Japan were the small alluvial lowlands, highly dispersed and separated from one another by hilly or mountainous country; this would have favoured the survival of powerful local clans and hindered the growth of a central authority.

With their superior agriculture the Japanese prospered, increased in numbers and spread northward to occupy all the best land in Honshu. By the end of the eighth century they had settled all the lowland as far north as modern Tokyo, displacing the sparse population of aboriginal Ainu who had only a primitive hunting economy. In northern Honshu the Japanese colonization was a slower process, for there they met cooler conditions which necessitated more adaptation—including some changes in their crops. By A.D. 1400 all Honshu had been occupied. Only Hokkaido remained as a frontier of settlement: even today the population density there is under 200 per square mile, compared with 800 per square mile for the remainder of Japan.

The seventh and eighth centuries were a fruitful period of Chinese contact. Japanese scholars and administrators visited China to study the achievements of the Tang dynasty. They adapted the Chinese written language to Japanese, which previously had no written form. A centralized political system headed by an emperor began to evolve during this period, but ultimate power remained with the feudal clan leaders until well after Tokyo became the capital in the seventeenth century.

By that stage the Japanese rulers had become suspicious that continued foreign contacts, not only with China but also with the Europeans who were beginning to appear around their coasts, might endanger their power. They decided on the complete isolation of Japan from the rest of the world, enforced by severe penalties on all who tried to enter or leave the country. No foreign contacts of any account were made for the next 250 years until the middle of the nineteenth century.

THE RISE OF MODERN JAPAN

The 250 years of isolation ended in 1853 when the Japanese under American pressure decided to resume commercial relationships with other countries. The Japanese objective was a thorough modernization of their country: students were sent to Europe and America to study industry, government and the armed forces; railways, ships and factories were built; ports were improved. By the turn of the century the Japanese could make practically all the industrial goods formerly produced only in such countries as Britain, Germany and U.S.A. The coastlands of the Inland Sea and the districts of Nagoya, Yokohama and Tokyo became the industrial zone of Japan. The city of Nagasaki and its surroundings in northern Kyushu became the principal area of iron and steel production, shipbuilding and other heavy engineering. The Osaka and Kobe areas have many of the textile and light-engineering works. Nagoya is a textile centre. Tokyo has a great variety of light industry. At the 1955 census Greater Tokyo had a population of 9.675 million. Osaka, Nagoya, Yokohama and Kobe all had over one million people.

Alone among the states of Monsoon Asia, Japan today has a range of industries as comprehensive as any to be found among the western countries. Japanese in the cities and larger towns patronize self-service and department stores, night-clubs and baseball games. Their personal equipment embraces cars, television, refrigerators and washing machines. These things are not only for the few; they are used by a high proportion of the urban population. The rural population, roughly half the total, is less well off. But their living standards even before World War II were substantially better than those of other farming areas of Monsoon Asia. The land reforms which followed the war brought great improvements. Small farmers who were previously oppressed tenants now own their holdings, even though a third of these are smaller than one acre, and farms over five acres are exceptional. Some farmers manage to buy modern domestic equipment for their homes and light machinery for their fields.

The country is poor in almost all industrial raw materials. There are coalfields in northern Kyushu, in eastern Honshu north of Tokyo and in Hokkaido, but the coal seams are mostly thin, broken and discontinuous. Ventilation and drainage are difficult in several mines which extend beneath the sea. Reserves are sufficient for the present annual output of around 50 million tons, mostly bituminous, to be maintained for many decades, but coking coal is very scarce. A few oilfields on the west coast of northern Honshu and in Hokkaido have contributed up to 15 per cent of the national oil requirements. Other fields might be discovered, but there is no indication of great reserves. Good use is made of water power:

about two thirds of the electricity produced is from hydro-electric stations. Among the minerals only copper exists in substantial quantities relative to national needs. Until World War II Japan was practically self-sufficient in copper, but in recent years home production has been only half the annual requirement. Rather less than half the manganese and magnesium used comes from Japanese mines.

Lack of capital, skilled labour and markets is regarded today as a major obstacle to industrialization in Monsoon Asia, yet the nineteenth-century Japanese and their twentieth-century successors achieved a comprehensive industrial development within a few decades, and that despite the poverty of their homeland in natural resources. Two questions arise: How did the Japanese do it? And if they could do it over 60 years ago why is it difficult for other Monsoon Asian states to do the same now?

The Japanese owe their economic success to their tremendous determination, hard work, social discipline and political and military ruthlessness. They were helped by the political and military weakness of neighbouring powers, which enabled them to dominate markets and raw materials in territory not their own; and also by the fact that the new industries and armed forces were controlled by a few powerful families for whom the rank and file were prepared to work at very low wages. The emperor formed the apex of this hierarchy, and was considered a divine authority. Great sacrifices in material living standards and personal contentment were accepted without question by successive generations. During the quiet period of isolation, moderate wealth which might otherwise have been dissipated in local wars accumulated in the hands of ruling feudal families and later became available for the new industries promoted by those same families, so that the problem of capital was overcome. Loans were raised in London and other financial centres. In copying western technology the Japanese rulers were careful to exclude western ideas of democracy which might have impaired their hold on the people.

Silk and cotton goods were the first factory products. For these, markets were found in America and Europe, and later in India. Goods made of cotton, silk and artificial fibres are today the largest single group by value of all Japanese exports. The raw silk is home-produced, but well before 1900 the population of Japan had risen to a level at which no land could be released from food production to grow cotton. India and U.S.A. became the main sources of cotton yarn and fibre. Other goods,

16. Malaya: a gravel-pump tin mine

17 A crowded quay on Singapore river, in the heart of the city

ranging from fountain-pens to bicycles and watches, were soon to be produced so cheaply that oversea markets were not hard to find. The population, which had remained for decades before 1850 at around 30 million, began to increase rapidly. The figure for 1900 was 45 million. From that stage onward food imports were essential to supplement supplies from Japanese farms: foreign food could be bought only with industrial exports; having started industrial development, the Japanese had therefore no choice but to continue.

As industries grew, dependence on imported raw materials increased. The advantage of oversea possessions as markets and as suppliers of food and raw materials was soon realized. Taiwan was acquired from the Chinese in the war of 1894-5 and Korea was annexed in 1910. The richest acquisition was that of Manchuria in 1931. There the Japanese built ports and railways, worked the coal and iron ore and supervised the immigrant Chinese settlers in the production of commercial crops (see page 195).

The world economic depression of the 1930's brought special problems for the Japanese. Manufacturing countries which had formerly admitted Japanese factory products sought to protect their own industries by reserving their home markets. The Japanese responded with a tougher foreign policy, envisaging the ultimate combination of China, Manchuria, Korea and as much as possible of the rest of Asia into one great economic enterprise which was to provide food, raw materials and markets for Japan. This was the concept which became known as *The Greater East Asia Co-prosperity Sphere*. In 1937 China was attacked. Encouraged by the German successes in North Africa and Europe during 1940 and 1941, Japan joined the axis powers. Attacks were mounted almost simultaneously on Hong Kong, Malaya and the American naval base at Pearl Harbour in the Hawaiian Islands. Lacking in aircraft and not properly prepared for jungle fighting, the small Commonwealth, Dutch and American defence forces had only a poor chance. Hong Kong surrendered on Christmas Day 1941 and Singapore six weeks later. Before the end of 1942 the Japanese had secured all South-east Asia except part of southern New Guinea. Their forward troops, supplied from bases in Burma, were moving into adjacent parts of Assam and Bengal. With all this they also gained an unsurpassed reputation for cruelty and contempt for the rules of civilized war.

If the war had been won by the axis powers this Greater East Asia

Co-prosperity Sphere, with a total population of perhaps 700 million, rich in industrial raw materials, would have become a reality. In the event, under the peace arrangements of 1945 the Japanese lost all but their homeland. Most of the conquered territory in South-east Asia was held for only two or three years but Taiwan and Korea were held for 50 years, Manchuria for 15 years and much of north China for nearly a decade. Yet comparatively few Japanese settled in the oversea possessions. Of those who did, most were in the armed forces and in urban occupations; hardly any went on the land. This suggests that the economic benefits of the growing co-prosperity sphere contributed substantially to maintaining living standards in the homeland at such a level that large-scale emigration remained unattractive to the Japanese, despite acute land hunger and overcrowding. All the oversea Japanese were repatriated after 1945 under the peace arrangements: they numbered only five million.

Sixty years ago the Japanese became the first Monsoon Asian state to industrialize and to launch a wide range of factory products on world markets. Their success was hard won, but they had the advantage of production costs lower than those of their competitors, the principal reason for these lower costs being the lower wages paid in Japan. A modern repetition of that Japanese success by other states of Monsoon Asia is impracticable. Today the provision of very large and efficient industrial plants aided by automation is generally more important than low wages in achieving low production costs. But such large industrial units, with finance and research to ensure their continued development, are mainly in U.S.A. and Europe. They are beyond the means of under-developed countries. Industrial development by such states as Ceylon, Malaya and Indonesia can now be achieved only in co-operation, not in competition, with western countries. Capital would almost certainly have to come from Europe, U.S.A. and the Commonwealth, though some might be forthcoming from Japan.

THE POST-WAR ECONOMIC PROBLEMS OF JAPAN

Foreign trade is vital to the Japanese. The population of Japan, 70 million in 1937, is now over 90 million and will probably reach 107 million by 1990. Despite the contributions of the sea fisheries, and also the great efforts of the farmers to increase their output, food production is unlikely to exceed 75 per cent of the national requirement even in good years. In addition to a quarter of the food supply, many industrial raw materials must be bought oversea. For example, 80 per cent of the iron ore and salt, practically all the coking coal and crude oil, and all the cotton, rubber and bauxite are imported. Before World War II the foreign trade of Japan became adjusted to the relatively favourable conditions of the developing co-prosperity plan. After the war the new political régime in China and the loss of the conquered territories presented acute difficulties which the Japanese survived only with American economic help. Japanese trade is still in the process of adjustment to post-war conditions. Some of the problems are indicated by the figures in Table 7. Before the

TABLE 7

JAPAN: DISTRIBUTION OF FOREIGN TRADE

(percentages)

	1934–6 (average)	1952–60 (average)
Imports from		
Mainland China	10	2
Taiwan and Korea	27	2
South-east Asia, Indo-Pakistan and Ceylon	16	22
North America	25	45
Europe	10	8
Oceania	7	8
Exports to		
Mainland China	18	1
Taiwan and Korea	25	9
South-east Asia, Indo-Pakistan and Ceylon	18	30
North America	17	25
Europe	8	11
Oceania	3	3

war Taiwan, Korea and North America together supplied over half the imports and took nearly half the exports. South-east Asia, Indo-Pakistan and Ceylon supplied 16 per cent of the imports and absorbed 18 per cent of the exports. The figures indicate an adverse trade balance with North America (imports 25 per cent, exports 17 per cent) and a credit balance with mainland China (imports 10 per cent, exports 18 per cent). The picture is very different for the years 1952–60. Trade with mainland China, Taiwan and Korea all but ceased. The export figure of 9 per cent to Taiwan and Korea includes military and other stores sent to Korea during and after the hostilities of 1951. South-east Asia, Indo-Pakistan and Ceylon together account for an increased share of both imports and exports. Two striking changes are the reliance on North America alone for nearly half the imports and the adverse trade balance with that area. Such an adverse balance was of little account before the war, when national currencies were freely convertible, but is serious today because credits earned in various Asian currencies are not convertible. Profits of trade with the poorer Asian countries are generally in currencies unacceptable to the 'hard currency' countries such as U.S.A.

The obvious and most appropriate solution to the Japanese economic problem is more trade with mainland China and less dependence on the U.S.A. The Japanese have the technical skill to produce the equipment needed by the Chinese and other peoples of Monsoon Asia in their efforts to industrialize. It remains to be seen whether a political arrangement can be made with the Communist government in Peking to open Chinese markets to Japanese and other foreign products. Such a provision might not solve all the Japanese difficulties; other countries would be strong competitors. Much of Japanese industry, though progressive in many ways, is still organized in small-scale units. Half the workers are in factories employing less than 50 people, and only 15 per cent in works employing over 100. The corresponding figures for U.S.A. are 16 per cent and 32 per cent respectively; and for Britain 26 per cent and 28 per cent respectively. Thus the greater proportion of American and British industry organized in larger units makes for efficiency of production which could outweigh the Japanese advantages of relatively low wages and proximity to the Asian markets (see also page 215).

KOREA

The physical setting. Korea is described by Professor C. A. Fisher as 'a land bridge, projecting some 600 miles from the parent continental mass to within 120 miles or so of the Japanese archipelago'. The peninsula itself, about 400 miles from north to south, has a maximum width of 200 miles. A mountain range with large areas over 3,000 ft. runs parallel to the east coast. The main rivers drain westward through extensive lowlands, but the lowland on the east coast forms only a narrow belt. There are no good natural harbours on the east, though protected inlets abound on the south and west coasts. Across the root of the peninsula near the border with Manchuria and U.S.S.R. the mountains rise above 6,000 ft. There is an easy lowland passage in the west into Manchuria and a more difficult coastal route on the east into U.S.S.R. The present political border follows the Yalu and Tumen rivers. In the southern half of the peninsula the annual rainfall is between 40 and 80 in., depending on exposure to the summer monsoon; summer temperatures are high, winters are cold, but the ports are usually free of ice. In the north the rainfall is generally less than 40 in., winters are severe and the ports and rivers are frozen for two to three months.

Minerals. There are useful reserves of iron ore in the lowland zone between Seoul and Pyongyang. The main coalfields lie south-east of Seoul, and there are others in the Pyongyang area. Other minerals well represented are tungsten, graphite, copper and gold: in all except gold the north is vastly richer than the south.

Historical geography. Two thousand years ago the northern half of the country was part of the Han empire. From the sixth to the sixteenth century the Koreans were an independent nation. Their culture, as that of Japan, owed much to China. In 1627 the peninsula was invaded by the Manchus, who were anxious to have a secure eastern flank for their advance into China. Korea remained a tributary state of the Manchu régime until the closing years of the nineteenth century.

Korea in the twentieth century. Surrounded by Japan, Russia and China, Korea held special significance for all three. To the Japanese, a Korea dominated by either Russia or China constituted a base from which the main islands of Japan might be threatened. The Japanese also saw Korea as an under-populated country with scope for development in agriculture and mining, and as a base from which they could begin their domination of the mainland. To the Russians the value of the peninsula lay in its ice-free harbours. As for the Chinese, they were (and are) traditionally reluctant to abandon claims to territory their forbears once held.

In 1896 a Japanese proposal to divide the peninsula along the 38th parallel into a northern protectorate supervised by Russia and a southern one controlled by Japan was rejected by the Russians. The defeat of Russia in the Russo-Japanese war of 1904 brought international recognition of all Korea as a Japanese protectorate. Formal annexation by the Japanese followed in 1910. Thereafter the Japanese ruled with brutality. Only half the cultivable area had been worked by the Koreans. Agriculture was extended and improved so that the country became the most important single source of Japanese food imports. Harbours were built. New railways opened up the mineralized area and made easy the rapid deployment of Japanese troops along the Russian and Chinese borders. Among industrial raw materials provided for Japan were mineral ores, cotton, silk, and soya beans.

After 1931 economic developments in Manchuria and Korea were co-ordinated. Impressive hydro-electric schemes on the Yalu river provided power for the heavy industries in the towns south of Mukden. Apart from a small iron and steel works at Seoul there was not much industrial development in Korea itself. The Koreans had a miserable time. Their living standards were among the lowest in Asia. Some were attracted by the lowest-paid jobs in Japan; others took part in settlement schemes in Manchuria, where their experience of poverty enabled them to outdo the Chinese in putting up with harsh conditions.

At the end of World War II the allied powers envisaged a unified and independent Korea, and agreed on a temporary division of the country along the 38th parallel into a northern portion to be supervised by Russia and a southern one to be supervised by U.S.A. The U.S.A. devoted immense funds to the restoration and improvement of agriculture in South Korea, while the north became the Korean People's Democratic Republic under strong Soviet and Chinese influence.

In June 1950 North Korean forces attacked South Korea. This brought an immediate response from President Truman who sent an American force to stop the aggression. Truman risked massive Russian participation on the side of North Korea, and probability of another world war. In the event the Americans were shortly to be supported by other United Nations forces, and the Russians held back. Though several nations made notable contributions to the fighting on the United Nations side, the Americans carried the main burden. In the early stages they were greatly outnumbered and were all but thrown out of the peninsula. Later the North Koreans were driven back across the 38th parallel. As the Yalu river border was approached, the Chinese intervened on the side of the North Koreans. Ultimately a cease-fire line near the 38th parallel was accepted by both sides. The civil population had suffered severely in the fighting and most of the towns were destroyed. In South Korea, which now has a defence treaty with the U.S.A., American help has been con- tinued. The agricultural improvements sponsored by the Americans in the south, coupled with the mineral wealth which is mainly in the north, would have provided a sound economic basis for a united Korea. But the achievement of that aim is prevented by the strategic value of the penin- sula to both the American and Sino-Soviet power blocs. Of the total population of around 36 million, about 25 million are in South Korea, where climatic conditions offer better scope for agriculture than in the north.

Additional note on Japanese export trade (October 1963)

The idea that the Japanese might ease some of their economic diffi- culties by more trade with Monsoon Asian countries and less dependence on the U.S.A. is attractive to western countries because such a policy might reduce Japanese competition in western markets. But the Japanese clearly expect that the purchasing power of Monsoon Asian countries will not grow fast enough to constitute an adequate market. In their official target, as published recently by the Fuji Bank of Tokyo, the Japanese envisage a three-fold increase in the value of their total exports by 1970 over the 1959 figure: but the *proportion* of the total exports destined for the U.S.A. and Europe in 1970 is expected to be about 33 per cent above that for 1959, with a reduction of about 33 per cent in the *proportion* going to Asian countries. It will be interesting to see if western countries can increase their purchasing power sufficiently to sustain their own industries and also absorb a growing supply of Japanese products.

15

South-east Asia

South-east Asia consists of the mainland lying to the east of India and south of China, together with the islands to the south and east, including the Philippines and western New Guinea. All but the north-eastern fringe of this area is inter-tropical.

On the mainland the most extensive alluvial lowlands are along the lower and middle courses of the Irrawaddy, Sittang, Chao Praya, Mekong and Red rivers. The sixth great river of the area, the Salween, flows for over 1,300 miles in a deep and narrow valley: its lowland tract is limited to a small area near its mouth in the neighbourhood of Moulmein. All these lowlands are good rice lands. But only on the Red river alluvium are rural densities of population comparable with those of Bengal and the plains of China. For the other lowlands 400 per square mile is a representative figure. Moreover, until a little over a century ago the Irrawaddy and Mekong deltas and the lower Chao Praya lands were empty swamps and forests similar to those seen today over large areas of northern Sumatra. Colonization by paddy farmers began only in the middle of the nineteenth century: on the Mekong delta a good deal of potentially productive land has yet to be reclaimed. The explanation of this delayed settlement lies partly in the physical geography and partly in the cultures of the local people. All the lowlands are easily accessible from the sea at their southern ends. The Red river lowland is also easily reached by a coastal route from the crowded areas of south China. The others are screened from both China and India by high, rugged and thickly forested country, and separated from one another by extensive uplands also largely forested. There was no pressure of population farther south which might have led to a northward migration from the island zone into the open ends of the lowlands, and the few people who penetrated the landward barriers from the north came with agricultural methods best suited to conditions drier than those of the deltas. For example, the ancestors of the modern Burmese entered the Irrawaddy drainage from the north. They found in the rain-shadow country around the Chindwin-Irrawaddy confluence a dry environment to which their farming traditions could be adapted. That area became the homeland of Burmese culture. The wetter country farther south demanded more adaptation than the early Burmese were prepared to make, and therefore remained repellent until about 1850 when the opportunities in the delta for rice production for export were realized.

The modern political divisions with boundaries for the most part fixed and internationally recognized date from the nineteenth century. Burma, Malaya and northern Borneo came under British supervision; the rest of Borneo, together with Java, Sumatra and many other islands, became the Netherlands East Indies, while the French established colonial rule in Indo-China. Only Thailand (Siam) remained a sovereign state, and that mainly because the British and French found it convenient to have their spheres of interest separated by a country free of European control and unlikely to be troublesome. The Philippines, Spanish territory from 1565 to 1898, were ruled not directly from Spain but from Spanish colonies in North America: their association with the New World continued under American domination until full independence was achieved in 1946. These modern political divisions do not coincide with racial or even linguistic or religious groups. For example, in Burma the Burmese are a majority, but the Shans, Thais and Karens are distinct cultural groups. And those groups are not wholly within Burma. They constitute local majorities over large areas astride the political borders with Thailand. Similar conditions prevail in Indo-China.

The last 20 years have seen the progressive replacement of foreign rule by independent local administrations, with boundaries practically identical with those of the colonial days except in the instance of Indo-China which was broken up to form the three states of Vietnam, Cambodia and Laos. The Netherlands retained control of western New Guinea until 1962, and part of Timor remains Portuguese. The only remaining British possessions are Sarawak, Brunei and North Borneo.

The Chinese in South-east Asia. Of a total population of around 215 million, about 10 million are Chinese in the sense that they or their ancestors came from China. A mere 5 per cent of the population, these

Chinese have had an economic and social importance disproportionate to their numbers for the reason that they dominated commercial activities such as the rice trade. Many are urban shopkeepers, craftsmen and market gardeners. Others occupy professional and clerical posts. A few are millionaires. Hardly any are peasant farmers. Except in the Philippines, the Chinese of South-east Asia have not intermarried with the other elements of the population. They form distinct social groups, for the most part more prosperous than the rest of the people.

The beginnings of Chinese contact with South-east Asia date from the Han dynasty, but the big influx of Chinese came in the late nineteenth century with the spread of tin-mining in Malaya and the colonization of the delta lands by rice farmers. Neither hard work for low wages in tin mines and plantations, nor the profits of urban commerce, held much attraction for the local peasants; industrious and thrifty Chinese came in thousands. At first their intentions were to remain only to accumulate savings before returning to China, but as the years passed an increasing proportion became permanent residents. Only very few Chinese now living in South-east Asia were born in China.

The Chinese number about 2,600,000 in the Federation of Malaya, just over one million in Singapore, where they form a 75 per cent majority, some three million in Indonesia and British Borneo and one million in Indo-China. The Peking government says there are three million Chinese in Thailand, though the Bangkok authorities give the figure as roughly one million.

The economic prospect. South-east Asia as a whole is not well off in natural resources which could be the basis of manufacturing industry on a grand scale. Locally there is scope for more small-scale industry. Rice, plantation products and tin are the traditional exports on which the area must depend in the effort to secure better living standards, but the outlook for those commodities is not entirely favourable. Though many Asian people go hungry, surplus rice is difficult to sell unless the price is low. Natural rubber, the main plantation product, can hold its place in world markets in competition with synthetic rubber only so long as production costs are kept down. Tin is so plentiful that its production has been restricted by international agreement. If costs are to be kept low, production and marketing must be efficient, and that in turn is practicable only where there is stable and efficient government, as for example in the Federation of Malaya. In several of the new independent states costs have tended to rise on account of demands for higher wages, and production has been hindered by the failure of the governments to maintain public order. For example, substantial parts of Indonesia and Burma have been dominated for long periods by armed forces in rebellion against the Jakarta and Rangoon authorities.

Another aspect of emergent nationalism has been the oppression of people best qualified to help some of the new states. Dutch nationals who supervised plantations and commerce were expelled by the Indonesians before local people were sufficiently trained to take their places. Special difficulties have arisen over the local Chinese in several countries. For the most part these Chinese are single-minded in their business activities and have no interest in the politics of China. But in China itself the traditional attitude has been that all Chinese oversea owe allegiance to the homeland; fear of the growing power of China, coupled with envy of the economic success of the local Chinese, has led to discriminative legislation—notably in Indonesia and Vietnam.

Unlike the remainder of Monsoon Asia, South-east Asia *as a whole* has no urgent land problem. Pressure of population is extreme only in Java, the Red river lowland, parts of the Philippines and on Singapore island. Elsewhere there are generally unoccupied tracts of forest and swamp which could be brought into cultivation if population pressure in adjacent farmlands became excessive.

BURMA

Flying eastward into Burma from the crowded plains of Bengal, air passengers are impressed by the great and uninterrupted extent of the northern Burmese forests, and by the Buddhist pagodas sited in or near the villages. The pagodas are graceful and architecturally pleasing, especially when built on a rock rising from paddy fields and backed by dense green jungle.

The eight principal regional divisions are shown in Figure 42.

The Dry Zone, the rain-shadow area to the east of the Arakan ranges, is mostly lowland. Everywhere the mean annual rainfall is less than 40 in., and in places it falls to 25 in. Some of the soils are sandy and red or

purple in colour. Others, especially those around the little river port of
Monywa on the Chindwin, are heavy, black and moisture-retaining. It
was in this relatively arid country that the Burmese culture first evolved.
Today it is an area of mixed farming. By Monsoon Asian standards
holdings are large: 10 to 15 acres is a common size. Rural population
densities are between 100 and 200 per square mile. Millet, cotton (especially on the black soils) and oilseeds are the chief crops. Rice is not
prominent: a little is grown wherever small canal systems provide
irrigation. Cattle, including water-buffalo, are raised in sufficient numbers
to maintain a supply of draught animals to the wetter areas farther south.

Yenangyaung, Chauk and Yenangyat are riverside settlements on the
oilfields which flourished under the direction of the Burma Oil Company before World War II. In those days crude oil was taken by pipeline
to refineries downstream from Rangoon. All the oil installations were
damaged during the Japanese invasion; the pipeline now remains derelict,
but the oil wells have been restored to give a small output sufficient for
national requirements. The crude oil is taken by river to Rangoon for
refining. In the south at Thayetmyo a cement works using local limestone
is fired by natural gas.

Mandalay, a former capital of the Burmese, had a population of
182,000 in 1955. An administrative centre for middle and upper Burma,
the town has various minor industries. It is a river port; that is to say, it
has a frontage on the Irrawaddy. As is usual at river ports in Monsoon
Asia, there are no quays. Steamers tie up at barges which are moored in
deep water and connected to the river bank by makeshift arrangements
of planks.

The sharp westward bend of the Irrawaddy south of Mandalay
indicates river capture on a grand scale. The beheaded Sittang is a misfit
in its wide valley. The Sittang route was chosen for the Mandalay-
Rangoon railway as an easier alternative to a line along the Irrawaddy
where badland country adjacent to the river presented engineering
problems.

The Irrawaddy and Sittang rice lands. Here the mean annual rainfall
increases from around 60 in. on the northern fringes to more than 100 in.
in the south and south-east. In the lower Irrawaddy and the delta rice is
practically the only field crop. One rice crop only is grown each year.
It is planted out from seed beds in July, soon after the early monsoon rain

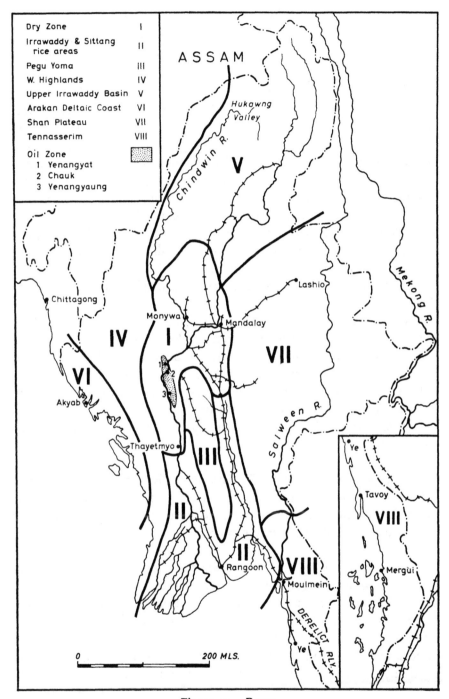

Legend:

Dry Zone — I
Irrawaddy & Sittang rice areas — II
Pegu Yoma — III
W. Highlands — IV
Upper Irrawaddy Basin — V
Arakan Deltaic Coast — VI
Shan Plateau — VII
Tennasserim — VIII

Oil Zone
1 Yenangyat
2 Chauk
3 Yenangyaung

ASSAM

Hukawng Valley

Chindwin R.

V

Mekong R.

Lashio

Chittagong

Monywa

Mandalay

I

VII

IV

VI

Akyab

Salween R.

Thayetmyo

III

II

II

VIII

Rangoon

Moulmein

DERELICT RLY.

Ye

0   200 MLS.

Ye

Tavoy

VIII

Mergui

Figure 42. *Burma*

has moistened the soil enough for ploughing: rain water accumulates in the fields as the rice grows. No river water is used; by December the mature rice is standing in dry earth ready for harvest.

The farms of the late-nineteenth-century Burmese colonists were small. As more land was reclaimed, thousands of Indian labourers were employed. The Burmese, unused to commercial farming, were easy prey to Indian money-lenders to whom they pledged their land for credit. Many Burmese failed to repay, and their holdings became merged into large estates controlled by absentee landlords; by 1939 most of the delta population were landless labourers. Post-war land reforms have gone a long way to restoring the rights of the local people. Jute, grown to provide sacks for the rice, is a recent introduction. In the Sittang valley, also, rice is dominant, but other crops are prominent locally. Many farmers do well out of sugar cane; others grow cotton.

Rural population density reaches 500 per square mile in the Irrawaddy delta.

Rangoon, the capital and only port of consequence, is on the side of the delta about 20 miles from the open sea. Accommodation for shipping is limited to quays on the city waterfront; there are no docks. Rice and timber milling, oil-refining and small-scale engineering are the chief industries. The population in 1955 was 737,000. Twenty years ago Rangoon was a staging post on world air routes, but this prominence was lost to Calcutta and Bangkok as the speed and range of commercial aircraft increased.

The Pegu Yoma, of folded Tertiary rocks, is hilly country generally below 1,000 ft. but rising to 5,000 ft. on the volcanic Mount Popa in the north. Lumbering has made great inroads on the teak forests. Soils are poor and the population thinly scattered.

The western highlands, which include the Chin Hills and the Arakan Yoma, are parallel ranges built of Tertiary rocks. They reach 12,000 ft. in the north; southward they become lower, meeting the sea as low hills in Cape Negrais. All this country is densely forested: the wetter, western slopes carry tropical rain forest; teak is cut in the deciduous forests of the drier eastern slopes and is taken by river to Rangoon. Shifting cultivators are thinly distributed in the hills. There are a few paddy farmers in valley bottoms.

The upper Irrawaddy basin. This enormous tract extending northward from the Dry Zone is thinly populated. The forests are rich in teak, and in the south there are scattered areas of rice farming. Altitudes rise to 20,000 ft. on the border with China. In the Hukawng valley at the head-waters of the Chindwin is a large lowland which could be cleared for agriculture if pressure of population in southern Burma became excessive.

The Arakan deltaic coast (Figure 43) consists of small alluvial tracts separated by forested ranges. The valley bottoms are rich paddy lands. Surplus rice is exported from Akyab. There are no all-weather roads to the Irrawaddy lowland. Access from the north along the coast is easy, and there have always been close cultural associations with Muslim Bengal.

The Shan plateau, lying mostly between 2,000 and 4,000 ft., has a sparse population of shifting cultivators. Locally, in the warmer lowland parts, higher population densities occur where sedentary farmers grow paddy. There is much open savannah country, and also forest; under a suitable farming economy the area could support many more than the present population. There are lead mines at Bawdwin. A few plantations produce tung oil and coffee. The road from the railhead at Lashio across the steep ranges which separate the headwaters of the Salween and Mekong leads to Kunming in Yunnan.

Tennasserim, extending from the mouth of the Salween southward into the Kra Isthmus, is similar in appearance to the Arakan deltaic coast. Areas of densely settled paddy land are separated by ranges of hills, but here the hills are of older rocks related to the Sunda Platform. Granites and associated tin-bearing veins have been exposed to erosion, with the result that alluvial tin ore has accumulated along the lower river courses. Alluvial tin is worked around Tavoy and Mergui. Moulmein, near the mouth of the Salween, has a trade in teak, rice and rubber.

The geographical pattern constituted by these eight regional divisions is that of a fairly closely settled central zone on the Irrawaddy and Sittang lowlands, and a peripheral zone mostly though not entirely forested and sparsely populated. The Irrawaddy, navigable northward for 900 miles and for an additional 2,000 miles on the delta distributaries, is a great natural asset. Large motor-driven vessels and a variety of small sailing boats are

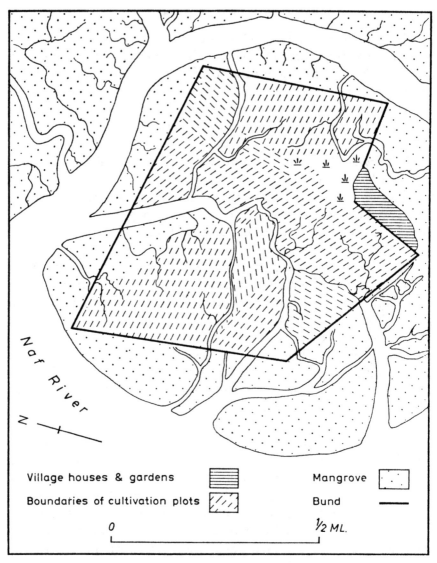

Figure 43. *Arakan coast: mangrove reclamation* This diagram was prepared from an aerial photograph. The bund is an earthen embankment to keep out tidal water. The old channels of the former swamp are clearly visible within the reclaimed tract; they now serve as drainage channels

the principal means of haulage. The metre-gauge railways supplement the river routes: they converge from outlying areas on Rangoon and Mandalay. Arakan and Tennasserim depend on coastal shipping for communication with Rangoon.

The estimated population of the country in 1962 was 22,342,000. Of the exports, the rice is about six times the combined value of the rubber, teak, tin and cotton.

MALAYA

The 11 regional divisions recognized by Professor E. H. G. Dobby are shown in Figure 44. Most of the population and economic activity are accounted for by five of those divisions: the tin and rubber belt, the northern coastal plain, the Kelantan delta, Penang and Singapore island. The remainder, amounting to 70 per cent of the total area, is forested country with very few people.

The tin and rubber belt. Alluvial tin is worked in valley bottoms; the rubber plantations are on higher, gently rolling ground. Both demanded good communications. The first railways were short lines, connecting inland mining areas with the river mouths at Port Swettenham and Port Weld which were accessible to coastal shipping through tortuous mangrove channels. The inland railheads were later connected by a line ultimately extended northward into Thailand and southward to Singapore. The towns of Ipoh, Taiping and Kuala Lumpur flourish as centres of mining districts. Chinese labour predominates in the tin workings. The plantation labour is mostly Indian, with some Chinese. Malay peasants also grow rice and small plots of rubber in this area, but they are a minority. Other commercial crops are oil palm, coconut and pineapple.

Alluvial tin is worked mainly by two methods, dredging and gravel-pump mining. Plate 14 is an aerial view of a tin dredge. The dredge floats in a shallow lagoon formed by local drainage into the excavation. The chain of steel buckets can be seen cutting into the working face on the left of the picture. The mud and gravel pass to the screening and washing plant amidships where the tin ore is extracted. The tailings, i.e. the waste material, is passed out at the other end of the dredge to fill part of the lagoon already worked over. At the gravel-pump mine in Plate 16 the site is kept clear of water by pumping. The working face of the alluvium is broken down by powerful water jets (monitors). The resultant slurry of ore-bearing mud flows down the long drain in the centre of the picture to be pumped up to the screening machinery at the far end of the working.

The northern coastal plain and the Kelantan delta are almost exclusively populated by Malay farmers. The main crop is rice, though there are small plots of rubber, coconut and vegetables. Rural population densities are higher than in any other part of the peninsula. The northern coastal plain has also a few rubber plantations staffed by Indian labour.

The west-central marshy coast—the 170-mile strip of mangrove and freshwater swamp forest up to 20 miles wide which hindered access to the tin and rubber belt until the railways were built—has few people. Port Weld, Telok Anson and Port Swettenham, the three ports with railway connections, are actually in the mangroves and require constant dredging. Malay communities on the seaward fringe fish and grow coconuts. Near Port Swettenham cultivated tracts of reclaimed mangrove surrounded by swamp indicate what could be achieved for the whole belt.

Georgetown on *Penang* island has the only deep-water anchorage on the west coast of Malaya. Chief British base in South-east Asia around 1800, it rapidly lost trade to the growing and more sheltered port of Singapore. Now it serves the north of the tin and rubber belt. Tin is smelted both on Penang and at nearby Butterworth and Prai on the mainland.

The east-coast fringe has a sparse population of Malay farming and fishing communities. Heavy seas during the northerly monsoon preclude fishing at that period, and the coastal people have to derive some of their food from ricefields.

The mountains and valley zone, the Trengganu highlands, the Pahang-Rompin valleys and east Johore constitute an enormous forested area of varied relief and very few people. Apart from a few Malay farmers on rare groups of paddy fields in valley bottoms, the only inhabitants are the Chinese and Indian workers who mine the iron ore in Trengganu and Johore and the tin near Kuantan. The laterized upland soils are unlikely to attract farmers, but there is plenty of marshy valley floor which could be drained for rice.

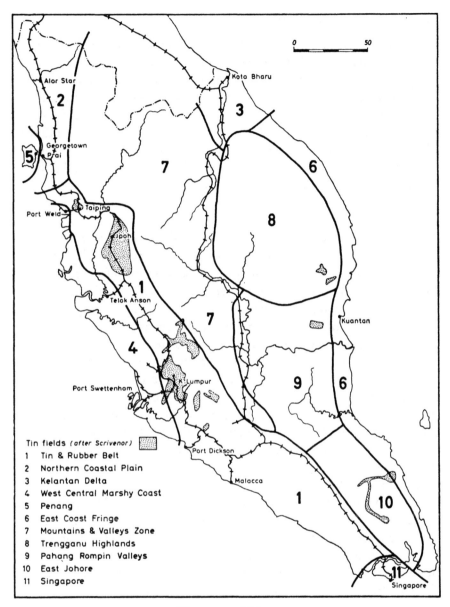

Tin fields *(after Scrivenor)*

1   Tin & Rubber Belt
2   Northern Coastal Plain
3   Kelantan Delta
4   West Central Marshy Coast
5   Penang
6   East Coast Fringe
7   Mountains & Valleys Zone
8   Trengganu Highlands
9   Pahang Rompin Valleys
10 East Johore
11 Singapore

Figure 44.   *Malaya*

Singapore was a forested and almost uninhabited island when Sir Stamford Raffles founded a British trading settlement there in 1819. It prospered as a commercial centre, especially with the extension of mining and plantation activities on the mainland; its banks and business agents attracted most of the entrepôt trade of South-east Asia. The docks are extensive and well equipped. There is a growing range of light industry. The population was estimated at 1,733,000 in 1962. The island is linked to the mainland by a granite causeway which carries the road and railway, and also a water pipeline from reservoirs in Johore.

Malaya: political geography

The political arrangement that evolved in the nineteenth century was complicated. Only Singapore, Malacca, Penang and a small part of the adjacent mainland were British colonial possessions. The remainder of Malaya consisted of states in treaty relationship with Britain. In 1957 the states and the colonial territory of the peninsula and Penang became a single independent sovereign power, the Federation of Malaya. The island of Singapore became an independent state in 1959.

The Federation prospers on the income of rubber and tin which contribute by value 70 per cent and 20 per cent respectively of the exports. The average life of the tin workings was estimated in 1946 at 24 years, but recently one large company claimed it could maintain its present production for 40 years. New reserves off-shore and in the valleys of eastern Malaya will probably be found and developed. The population of the Federation in 1961 was 7,137,000: 50 per cent were Malays, 37 per cent Chinese and 11 per cent Indians and Pakistanis.

In Singapore the economic prospect is uncertain. The population, 75 per cent Chinese, is increasing rapidly; this brings problems not only of living space and food supply but also of employment. Graduates of the university at Singapore, and others with good qualifications, vastly out-number the suitable jobs, and this condition will get worse. By 1972 half the population will be under 15 years of age and even larger numbers than at present will enter the labour market each year.

Singapore grew as the commercial focus for the whole of the peninsula of Malaya. Since 1957 the rulers of the Federation have tried to reduce their dependence on the services of the island. Port Swettenham, though still small, has been enlarged to handle more shipping. Import duties have been levied on Singapore factory products to encourage local industry. Some commercial companies have moved their regional head-quarters from Singapore to the federal capital, Kuala Lumpur. It is hard to visualize a solution to the employment problem of Singapore unless the island joins the Federation. While a large body of opinion in Singapore would welcome such a union, with its prospect of employment and economic opportunity throughout the peninsula, the Malays, who are now the ruling majority in the Federation, realize that if Singapore were incorporated the Chinese would be the most numerous racial group in the enlarged state. Political and economic control would quickly pass into Chinese hands.

THAILAND (SIAM)

Like Burma, Thailand has a central lowland drained southward by great rivers, hilly country with teak forests in the north, plateau country in the east and a peninsular area in the south (Figure 45). But there are important differences: Thailand is drier; all Thailand north of the peninsular part is a rain shadow relative to all summer monsoon air except that which comes directly from the south. Parts of the central lowland receive less than 40 in. of rain a year, and aggregates over 80 in. are known only in the peninsula and the high western borders. Also the hills and plateaux are not as high as those of Burma—the northern highlands reach only 5,000 ft. The Korat plateau in the east has a general level of 500 ft., though its upturned western edge reaches 1,500 ft.

Most of the people live in the southern half of the central lowland, where population densities are generally above 200 persons per square mile but locally exceed double that figure. These people are rice farmers, the majority being tenants working holdings on large estates. The rainfall is not sufficient for rice, so the crop is dependent on water distributed by canals from the Tachin and Chao Praya, which diverge south of Nakhaun Sawan to follow roughly parallel courses to the sea. The canals fill only when the river is in flood. The time and volume of the flood vary significantly from year to year. Until the barrage at Chainat was completed in 1957 there was no means of controlling the flow into the canals. In two years out of three large acreages were likely to be ruined because water was either insufficient or too plentiful. This trouble is being progressively overcome by new irrigation works.

In the northern half of the central lowland population densities are lower. Farming is mainly a subsistence occupation, with strains of rice tolerant of poor rainfall. Commercial crops such as tobacco, sugar cane and cotton are also raised. In the northern highlands teak is felled and sent out by the rivers, and also by the railway from Chiang Mai.

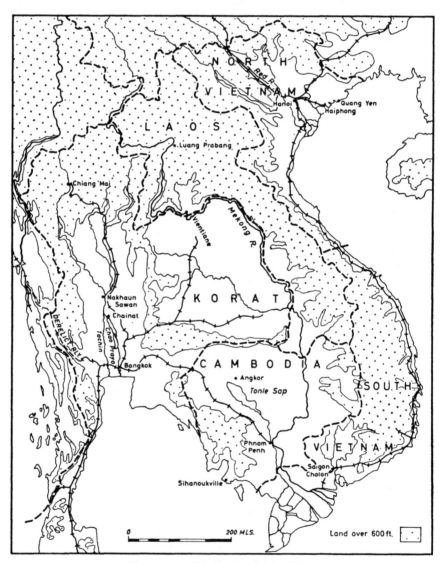

Figure 45. *Thailand and Indo-China*

The Korat plateau is built of sandstone through which water percolates easily. It is therefore drier in aspect than its rainfall figures might suggest. Population is scanty except in the rice lands along the rivers draining eastward to the Mekong. Along those rivers the rice is sown broadcast on the receding flood waters; no attempt is made to control the water level with embankments to enclose the fields—consequently large acreages of rice fail each year. Away from the rivers only shifting agriculture is practised. There is a local surplus of draught animals for export to the central lowlands.

The peninsular area of Thailand is similar to neighbouring parts of Burma: rice farming at the coast and on valley bottoms, and the hill ranges forested. Alluvial tin is worked with modern machinery and exported largely to U.S.A.

Of the exports, rice is by far the most valuable. Rubber, about one third the value of the rice, takes second place (the rubber estates are in the peninsula south of Bangkok). Tin and teak together amount to only half the rubber export. Communications are generally poor. Widespread flooding in the central lowland is an obstacle to road building. The Thailand rivers, in contrast with the Irrawaddy and Chindwin, are of little help to transport. Their depth and flow permit steamers to move inland from Bangkok about 250 miles, and that only during the flood season. Railways converge on Bangkok from Chiang Mai, the Korat plateau, Phnom-Penh in Cambodia, and from Malaya. The notorious Burma-Siam railway built by the Japanese with prisoner-of-war labour in 1942 is derelict.

Bangkok (population 2,300,000 in 1960) is 30 miles inland along the Chao Praya. It handles the entire export of rice, rubber and teak, but shipping facilities are poor.

INDO-CHINA

Indo-China was the collective name for the French colonial possessions in South-east Asia. Since the ending of French political control under the terms of the Geneva Conference of 1954, it has been a convenient geographical label for the combined areas of the three new independent states of Vietnam, Cambodia and Laos.

In the north are the lowland and delta of the Red river. In the south is a much larger lowland, that of the Mekong. Between the two is the Chaine Annamitique, reaching 7,000 ft. in the south and merging northward with a broad expanse of mountain and plateau of similar altitude. In the south-west the Cardamom Hills reach 4,000 ft.

Peasant farmers have sought out the best rice lands, namely the Red river lowland and delta, the Mekong lowland and delta, and small alluvial areas of the narrow and discontinuous coastal plain east of the Chaine Annamitique. The uplands are malarial and have a scanty population of shifting cultivators, except in the south of the Chaine Annamitique, where fertile soils derived from basalts support plantations of rubber and tea. On the Red river delta rural densities of population reach 1,500 per square mile. The Mekong delta was brought into cultivation more recently: there were few people there before 1850. Even today the rural density is only around 250 per square mile, mostly tenants on large estates as in the Chao Praya lands of Thailand. The main reasons for this late colonization of the Mekong delta are the comparative isolation from the north and the difficulty of controlling the distributaries on account of their enormous silt load and rapid rate of deposition. Upstream from the delta, the Mekong lowlands were the physical setting of advanced agricultural states in early times. Around A.D. 900 the Khmer rulers of Cambodia chose a site 150 miles from the Mekong for their new capital, Angkor. The ruined temples of Angkor near the great lake of Tonle Sap are evidence of the high cultural achievements of the tenth-century Cambodians, and also of strong Indian architectural and religious influence.

French efforts to restore their rule in Indo-China after the Japanese surrender of 1945 resulted in long and bitter fighting against local nationalist forces. At the Geneva Conference of 1954 the French agreed to relinquish all claims to political control, and the new states of Vietnam, Cambodia and Laos received international recognition. It was also agreed that, as a temporary measure, Vietnam should be divided into northern and southern portions at the 17th parallel until elections could decide whether the Communist régime which held the north or the democratic organization of the south should govern the whole state. The elections have not taken place. The division has hardened into a permanent arrangement: the north under Chinese Communist influence, the south progressing with the help of lavish American aid.

North Vietnam (population about 14 million in 1959) includes the Red river lowland and bordering hills, and a coastal area southward to the 17th parallel. The capital, Hanoi, has a population of about half a million. Near the coast north of Hanoi the Quang Yen coalfield produced nearly two million tons of anthracite in 1958.

South Vietnam (population about 13 million in 1959) includes the coast zone and the Chaine Annamitique south of the 17th parallel, together with the Mekong delta. The capital, Saigon, stands on a navigable waterway to the north of the delta. With its twin town of Cholon, it had a population of 1,300,000 in 1958. There is a large annual rice surplus for export.

Cambodia (population about five million in 1958) comprises the lowlands of the lower Mekong and the plains around the Tonle Sap, together with the forested Cardamom Hills. Population density on the lowland averages no more than 40 per square mile. The Tonle Sap shrinks in the dry season to about 1,000 square miles; in the wet season it receives flood water from the Mekong in addition to local water and extends to 4,000 square miles. Fish from the lake are an important source of food. The capital, Phnom Penh (population 500,000 in 1959), is linked by railway with Bangkok and by road with the new Cambodian port of Sihanoukville which was opened in 1960.

Laos (population 1,700,000 in 1959) is a backward, forested country on the upper Mekong, extending north through the hills and plateaux to the Chinese border.

The Mekong development scheme

Surveys began in 1958 for a scheme sponsored by United Nations for seven dams on the Mekong. At least three of these would be in Laos, and the scheme would bring electric power and controlled irrigation to all the Indo-China states except North Vietnam.

INDONESIA

During their period of colonial rule the Dutch devoted most of their attention to Java and its neighbouring small island of Madoera. The remainder of the Netherlands East Indies constituted the Outer Provinces, for the most part thinly populated and economically undeveloped, although there were small areas of intense plantation and mineral activity —as in Sumatra and Borneo. So that coffee, sugar, indigo and spices should be available for export, the Dutch in 1840 introduced into Java the 'culture and consignment' system. This meant that villages had to devote one fifth of their land to those commercial crops and were responsible for delivering the harvested crops to government agents. The arrangement worked well while there was plenty of land and labour for food production. But too often local Dutch officials demanded more than a fifth of the land to be given to commercial crops, and the subsistence crop acreage was correspondingly reduced. Abuses of that kind led to the failure of the system within 30 years, but while it operated the Javanese realized that there was an economic advantage in having large families to work in the fields. This fact, together with the natural richness of the soils derived from basic volcanic materials, encouraged a rapid growth in population in Java from about 4,600,000 in 1815 to 12,500,000 in 1860. After 1860 the culture and consignment system was progressively replaced by privately sponsored plantations, though it did not disappear entirely until 1918.

In 1961 Java and Madoera had an estimated population of 63,000,000 at an average density of 1,234 persons per square mile, compared with 32,655,000 at an average density of 48 for the Outer Provinces, where soils are generally poorer and less economic development has been attempted. In Java the Dutch built the most efficient railway system in South-east Asia. Outside Java there are practically no railways in Indonesia except in Sumatra. Several schemes for resettling people from overcrowded Java in the Outer Provinces were started by the Dutch and are continued by the Indonesian government; so far, though, comparatively few people have moved out of Java.

Of the exports of Indonesia as a whole, oil and rubber each account for 45 per cent by value and tin 4 per cent.

JAVA

A chain of volcanoes, 17 of them active, forms the highest ground rising to 12,000 ft. Between the volcanoes are several alluvial basins with floors between 2,000 ft. and 3,000 ft. North and south of the volcanic chain marls and limestones of Tertiary age form hilly country, and they in turn are succeeded by coastal lowlands built of alluvium, much of which derives from the volcanic material inland. The coastal lowland is well represented only in the north. In the south it is absent except near the port of Tjilatjap and a few localities farther east (Figure 46). The island of Madoera and the hills north of the Solo river are of Tertiary rocks, mainly limestone. Volcanic deposits, Tertiary rocks (marls and limestones together) and alluvial material each occupy about one third of the island. Rainfall in the western half of the island is high and well distributed

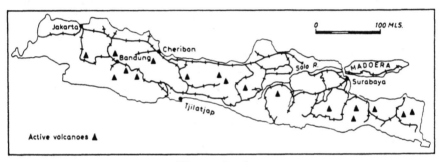

Figure 46. *Java*

throughout the year; in the east annual aggregates fall to 40 in. and there is a marked dry season between June and September.

The Tertiary rocks weather to poor soils. They are generally forested. In the east they carry valuable teak, which is worked especially in Madoera and the hills north of the Solo river. The alluvial basins between the volcanoes, and the lower slopes of the volcanoes themselves, have rich soils and support dense rural populations, the highest population densities occurring on the coastal lowlands in the north where 3,000 per square mile over large areas is common.

The northern lowlands to the west of Cheribon form two distinct surfaces. The lower surface occupies the coastal strip and is given almost entirely to paddy. Inland, the upper surface has equally good soils, but flooding of paddy fields is less easy and more land is under sugar cane and plantation crops. The slope between the two surfaces involves a sharp rise of around 60 ft. The railway from Jakarta to Cheribon runs along the crest. Jakarta (population about three million) and Surabaya are two large ports. Together with Cheribon, they have textile and light-engineering industries. There is a small oilfield near Surabaya.

SUMATRA

The island falls into three divisions (Figure 47): a mountain belt rises sharply from the west coast; on its eastern flank is a foothill zone of Tertiary rocks similar to those of Java; to the east again is a swampy alluvial lowland with a mangrove fringe.

The mountain belt, with ten active volcanoes, is more complicated in structure than the highland zone of Java. Volcanic materials take up a smaller proportion of the surface, and there are slates and shales as old as Carboniferous. Two parallel ranges can generally be discerned, with the depression between them blocked in places by volcanic accumulations. Several mountain basins with alluvial floors support dense populations; the slopes to the west coast are everywhere steep and forested. In Atjeh, the northernmost district of the island, the highland interior is only thinly populated. Large areas are under alang grass, thought to have established itself after the forest had been destroyed by shifting cultivators. Farther south the highlands lying between Sibolga on the west coast and Medan on the north have large tracts of good volcanic soils and a moderately dense settlement. Southward again, soils are generally poorer and there are fewer people, as far as the volcano of Kerintje; the slopes of Kerintje, and the valley between the two ranges that run southward to the Sunda Strait, have rich volcanic soils. Javanese have taken up land in that area in recent decades, and there is room for many more.

The coastal plain on the west is discontinuous and narrow: Padang with its port of Emmahaven is the only large settlement. Coal from the small Tertiary basin of Sawahloento, in the hills about 40 miles north of Padang, is used by coastal shipping and in a local cement works, but it is of poor quality and has a tendency to spontaneous combustion if left too long in the bunkers of ships.

The foothill zone of Tertiary rocks and poor soils is not attractive farming country. It has a few plantations, and provides firm ground for roads and railways along the edge of the swampy lowland. In the south, 80 miles up the navigable Moesi river from Palembang, is the small Tertiary coalfield of Bukit Asem. The coal is used on river boats and railways: it has a higher calorific value than the Sawahloento product, largely because it was heated and hardened at some time by adjacent volcanic material.

The alluvial lowland reaches a maximum width of 100 miles in the

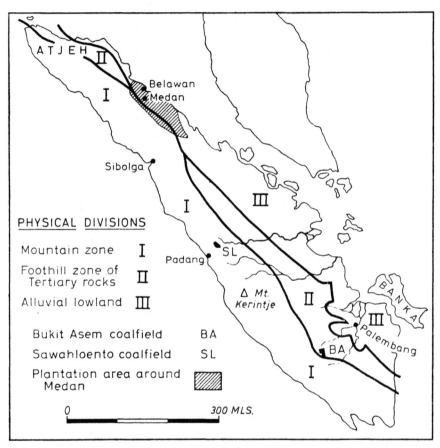

Figure 47. *Sumatra*

south. The narrow northern part has a dense population of paddy farmers: and the wider part in the south is mostly swamp; larger than the Irrawaddy delta, it is potentially rich farming land. Anticlinal structures beneath the alluvium yield oil in several places, notably around Palembang in the south, and Pangkalansoesoe, about 100 miles north of Medan (Figure 48). During the Dutch period large private funds were invested

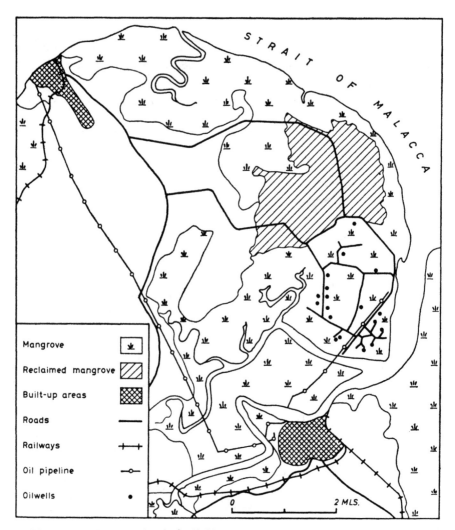

Figure 48. *Sumatra: detail of oilfield in mangrove swamp* The northern of the
two built-up areas is Pangkalansoesoe: the southern one is Pangkalanbrandan.
The oilwells in the swamp are served by roads, and by a pipeline leading
southward to Pangkalanbrandan where the sea approaches are too shallow for
all but small tankers. Another pipeline leads from storage tanks on the west
side of the town to Pangkalansoesoe where larger ships can come alongside.
North of the oilfield is a large tract of reclaimed mangrove given entirely to
paddy: the dwellings of the farmers are scattered along the road, and along the
low embankment which forms the boundary with the swamp. Land areas left
white are thinly populated and forested country all under 50 ft.

in plantations in the country around Medan. The soils of the hill country are volcanic, and those of the lowland are derived from the volcanic material inland. This plantation country (Figure 47) is remarkable for its range of products: tobacco, coconuts and oil palm on the lower ground; tobacco, rubber, tea and kapok at higher levels. Medan itself is 20 miles inland on the Deli river. There is a small modern port nearer the river mouth at Balawan.

The off-shore islands of Banka and Billiton, parts of the Sunda Platform, have alluvial tin which is sent to Penang for smelting.

KALIMANTAN (INDONESIAN BORNEO)

Borneo, the largest island of South-east Asia and part of the Sunda Platform, is hilly: several ranges exceed 7,000 ft. Apart from a few coastal tracts the whole island is forested, and except on the alluvial ground the soils are highly laterized. In the hills of the interior there is only a sparse population of hunters and shifting cultivators. Malaysian rice farmers have settled on the alluvium of the lower river courses, but there is still much potential rice land awaiting development.

Kalimantan, the Indonesian part of Borneo, had an estimated population of only 3,700,000 in 1958—mainly peasant farmers in the hinterlands of Singkawan, Pontianak and Bandjermasim. There are a few rubber and coconut plantations, but the most valuable product is the oil from the east-coast fields of Tarakan and Balikpapan. The Indonesian government is progressing with extensive swamp reclamation in the south, to relieve population pressure in Java.

BRITISH BORNEO

Here the relief and climate resemble those of Kalimantan. There are three political divisions: *North Borneo* (area 30,000 square miles, population 470,000); *Brunei* (area 2,230 square miles, population 87,000); and *Sarawak* (area 47,000 square miles, population 770,000). The smallest, Brunei, is a sultanate in treaty relationship with Britain. The other two are British colonies.

Taking the three divisions together, many of the population are Muslim farmers growing paddy and some rubber, coconut and pepper on coastal and riverine tracts; about 16 per cent are urban Chinese. The Dyaks of the interior, chiefly hunters and shifting cultivators, are a very small minority; there are a few large foreign-owned plantations, principally rubber. The area is dependent on substantial annual imports of rice, but there is plenty of good land awaiting reclamation and settlement. Internal communications are poor. Timber and plantation products are exported from such ports as Jesselton, Sandakan, Brunei and Kuching. Oil provides the principal wealth, but the Miri field in Sarawak is now nearly exhausted. Brunei has large reserves which earn increasing royalties, and interest on the invested royalties pays for the entire government expenditure.

It is remarkable that an area so rich in resources should have been so little developed during the last hundred years. The explanation lies partly in its distance from shipping routes, and the preoccupation of western powers with more accessible parts of South-east Asia—the oil is a comparatively recent discovery. Low population in relation to the area, and shortage of labour, are likely to remain handicaps to economic progress; but, given more people, there is plenty of scope in oil, plantation agriculture, forestry and the draining of swamps for paddy cultivation. Political union of the three territories—for which there has been much local and oversea support—would certainly help to realize these opportunities. As things are, some oil revenues of Brunei might be better spent. No one has reported Cadillacs in the jungle but the minaret of a new mosque has a powerful electric lift.

THE CONCEPT OF GREATER MALAYSIA

In 1961 the Prime Minister of the Federation of Malaya recommended that British Borneo and also possibly Singapore should join with the Federation to form a new political and economic entity, Greater Malaysia. Whether they want it or not, political independence is likely soon to be conferred on the three territories of British Borneo, and their best chance of progress would then almost certainly lie in some larger organization such as a Greater Malaysia. Malaya and Singapore together would have a Chinese majority unacceptable to the Malay population. The addition of British Borneo would bring over a million more people of whom less than a fifth are Chinese, and thus reduce the risk of Chinese domination.

This area, formerly part of the Netherlands East Indies, remained under Dutch colonial rule until 1962. The Indonesians call it Irian, or West Irian. In the highlands several peaks reach the snowline. Lowland swamps are extensive, especially in the south. The indigenous population of Papuan people numbers no more than 700,000. There is a small oil production, and deposits of nickel and cobalt have been proved. Near Merauke, Dutch engineers have reclaimed swamps for paddy cultivation. In 1962, under United Nations auspices, control of the area was transferred to the Jakarta government.

PHILIPPINES

Of the 7,000 islands and islets that comprise the Philippines, 11 account for 95 per cent of the land area. In his book *Southeast Asia* Professor E. H. G. Dobby suggests the following division as a convenient basis for geographical study:

1—Luzon, the well-populated island of the north;
2—the Visayan Islands, arranged roughly radially around the Visayan Sea;
3—Mindanao, the second-largest island, well to the south;
4—Palawan and the Sulu Islands, stepping-stones between the Philippines and Borneo.

All the islands have strong relief, some peaks reaching 9,000 ft. The most extensive lowlands are the Cagayan valley and the Manila plain of Luzon. Variations in exposure to the monsoons account for the climatic division of the islands into one zone which experiences a dry season in winter and spring, and another in which rain falls throughout the year. The zone with a dry season is a rain shadow relative to the winter monsoon: it includes most of Luzon, the western parts of Mindoro, Panay and Negros, and the western flanks of Palawan. Annual rainfall ranges from 40 in. at a few very sheltered stations to 180 in. on slopes exposed to the south-west.

The human geography is summed up by Dobby as 'congested pockets and plains separated by nearly empty forested interiors where wander aboriginal tribes or an occasional pioneer practises shifting cultivation'. About 15 per cent of the total area is cultivated. All the best land in Luzon is occupied, but in Mindanao there are good opportunities for further agricultural settlement. Rice, the main food crop, takes up 40 per cent of the cultivated area of the Philippines; maize accounts for 20 per cent, and root crops (mainly yams and cassava) about 4 per cent. Much of the land is double-cropped. About 25 per cent of the farmland is under commercial crops, predominated by coconuts; sugar cane and hemp are also prominent.

Though climatic conditions resemble those of Indonesia and Malaya, economic development in the Philippines has taken a different course from that of other parts of South-east Asia. The principal reason is that the Philippines are unique in South-east Asia for their long association with Central America and the U.S.A. Commercial agriculture was not encouraged by the Spanish. After the change to American colonial rule in 1898, sugar, fruit and tobacco cultivation was discouraged in order to protect the producers of those crops in Cuba, California, Hawaii and Georgia. By the time the Americans had pacified the islands the estates of Malaya and Indonesia had become so competitive that rubber had little prospect of success in the Philippines despite the favourable climate.

Of the exports, coconut products account for 40 per cent by value, sugar 30 per cent and timber 10 per cent. Iron ore and copper are sent to Japan. Rich new copper deposits have recently been found on Mindanao.

The Philippines are unique in Asia for the degree to which their indigenous and immigrant populations have intermarried. Spanish and Chinese have interbred with the Malay stock so that today there are no distinct racial groups.

16

Postscript

Fortunately for the geographer, his science does not require him to fore-tell the economic and political future of the regions he studies. His job is to explain the past and the present. The pitfalls of prediction are best left to others. But some reference to the future is appropriate at the end of a book of this kind because that future will rapidly become the present which the student will be required to explain. Political changes and economic trends which at the time of writing either cannot be discerned or seem too immature to claim much space in a textbook might well become great issues. So it is clearly essential that students keep in touch with current developments. Good sources of information are the news items and occasional special articles in *The Times* and *The Economist*, both of which present up-to-date facts. In general they do not give a geographical interpretation of these facts: that task is left to the student, with the help of textbooks.

Topics which might acquire increasing importance are:

1—*Political changes in India resulting in the establishment of new states based on considerations of language, religion and community;*

2—*Disputes between India and Pakistan over water supplies consequent on irrigation development in Bengal;*

3—*Chinese attempts at economic and political domination in South-east Asia;*

4—*Japanese pressure for trade with China;*

5—*Greater Malaysia (see page 240);*

6—*The Mekong development scheme (see page 233).*

It is certain that as and when the present problems of food supply, agricultural productivity and industrial progress in Monsoon Asia are solved, new economic problems will appear. For example, one consequence of simultaneous improvement in the farmers' lot and the establishment of more manufacturing industry in the towns might be that the farmers could afford to consume an increasing proportion of the food they produce—at a time when demand from the urban population is rising. This could result in an increase in food prices sufficient to threaten the political stability of a country, despite higher productivity on the land.

However, while educated Asians throughout the region recognize the urgent need for economic progress, they do not necessarily wish to achieve the high degree of material prosperity found in western countries. Asian people have their own values. Present indications suggest that they are unlikely to abandon leisurely ways in order to acquire as status-symbols the kind of personal effects for which people of western countries work hard and long.

Selected Bibliography

A disproportionately large amount of the world's best geographical literature is about Asia, and much of it about Monsoon Asia. The items in the list below are a small selection: most of them contain full bibliographies which are recommended to students seeking additional reading.

A. D. C. Peterson, *The Far East* (3rd ed. 1957). This is probably the best short introduction to the subject.

W. G. East and O. H. K. Spate (editors), *The Changing Map of Asia* (4th ed. 1961). The introductory chapter and Spate's contribution on India and Pakistan are especially recommended.

J. E. Spencer, *Asia East by South* (1954). A good textbook by an American geographer.

L. D. Stamp, *Asia* (11th ed. 1962). A comprehensive and useful reference work.

E. H. G. Dobby, *Monsoon Asia* (1961). A detailed and well-illustrated interpretation of selected areas and aspects of Monsoon Asia.

Other useful material, well presented, will be found in the following:

W. G. East and A. E. Moodie (editors), *The Changing World: Studies in Political Geography* (1956).

K. Pelzer, *Pioneer Settlement in the Asiatic Tropics* (1945).

Economic Survey of Asia and the Far East. An annual report published by United Nations.

A. B. Mountjoy, *Industrialization and Under-Developed Countries* (1963). This book is essential reading.

India and Pakistan

The book of outstanding quality, and one which owes much to the first-hand experience of the author, is *India and Pakistan* by O. H. K. Spate (2nd ed. 1957). Other recommended texts are:

Nafis Ahmad, *An Economic Geography of East Pakistan* (1958). A survey by the Professor of Geography at the University of Dacca, well illustrated by full-page maps.

H. C. Hart, *New India's Rivers* (1956). A survey of the multi-purpose hydro-electric schemes.

A. J. Coale and E. M. Hoover, *Population Growth and Economic Development in Low-income Countries* (1959). This is a study of India's economic prospects.

The following are suggested as initial reading on the historical background:

Percival Spear (editor), *The Oxford History of India* (3rd ed. 1958).

Philip Woodruff, *The Men Who Ruled India: The Founders* (1953).

W. H. Beveridge (Lord Beveridge), *India Called Them* (1947).

Lord Macaulay, *Essays*, especially the essay on Clive.

E. M. Forster, *A Passage to India* (1932).

John Connell, *Auchinleck* (1959). The biography of Field Marshal Sir Claude Auchinleck, who was Commander-in-Chief, India, at the time of the Partition of 1947.

Ceylon

E. K. Cook, *Ceylon* (2nd. ed., revised by K. Kularatnam, 1951).

B. H. Farmer, *Ceylon. A Divided Nation* (1963).

B. H. Farmer, *Pioneer Peasant Colonization in Ceylon* (1957). The same author contributed the chapter on Ceylon in Spate's *India and Pakistan*.

Leonard Woolf, *Growing* (1961). Colonial administrator, 1904–11.

China

G. B. Cressey, *Land of the 500 Million* (1955).

Theodore Shabad, *China's Changing Map* (1956).

Owen Lattimore, *Inner Asian Frontiers of China* (1940).

Owen Lattimore, *Pivot of Asia* (1950).

S. G. Davis, *Hong Kong* (1949).

T. R. Tregear, *Land Use in Hong Kong and the New Territories* (The World Land Use Survey, Monograph 1, edited by Professor L. D. Stamp, 1958).

T. R. Tregear and L. Berry, *The Development of Hong Kong and Kowloon as told in Maps* (Hong Kong University Press, 1959).

The following are excellent travel books:

Mildred Cable and Francesca French, *The Gobi Desert* (1942).

H. W. Tilman, *China to Chitral* (1951).

Japan

G. T. Trewartha, *Japan, A physical, cultural and regional geography* (1945).

E. A. Ackerman, *Japan's Natural Resources and their relation to Japan's Economic Future* (Chicago, 1953).

F. C. Jones, *Hokkaido* (1958).

South-east Asia

E. H. G. Dobby, *Southeast Asia* (6th ed., 1958).

Charles Robequain (trans. by E. D. Laborde), *Malaya, Indonesia, Borneo and the Philippines* (2nd ed., 1958).

C. A. Fisher, *South-east Asia* (1964).

W. Hodder, *Man in Malaya* (1959).

Ooi Jin-Bee, *Land, People and Economy in Malaya* (1963).

V. Purcell, *The Chinese in South-east Asia* (1951).

Colonel F. Spencer Chapman, 'Travels in Japanese-occupied Malaya', *Geographical Journal* (Vol. CX, 1947). An account of first-hand experiences in guerilla operations behind the Japanese lines in 1942–3.

Chapman's *The Jungle is Neutral* (1949) is also recommended.

Index

ADAM'S Peak, 149
Afghanistan, 23, 38, 56, 58, 61, 135, 136, 138, 139
aghani harvest, 114
Agra, 61, 62, 85, 86, 105, 112
Ahmad, Professor N., 139, 142
Ahmadabad, 86, 94, 119
Ainu, 206
Akbar, 61, 112
Aksu, 199
Akyab, 39–44, 221, 223
Ala Shan, 28
alang grass, 48
Alexander the Great, 57, 135, 136, 138
Allahabad, 39–44, 104, 109, 112
Alor Star, 39–44, 68, 227
alpine fold belt, 22–4, 29, 31
Altai mountains, 28
Altyn Tagh, 24, 30, 190, 191
Alwaye, 120
aman crops, 121, 122, 141
Ambala, 104, 110
Amoy, 188
Amritsar, 104, 109, 110
Andaman Islands, 26
Angkor, 67, 230, 232
Annam, 66
Anshan, 166, 170, 171, 182, 195
anthracite, 83, 233
antimony, 170
Anuradhapura, 150, 160
Arabs, 68, 85
Arakan, 26, 223–5
Arakan Yoma, 26, 48, 219, 222, 223
Aral Sea, 58
Aravalli mountains, 105, 114–16
Arial Khan, 139, 140
Arka Tagh, 24
Arun river, 24
Asansol, 95, 96, 102, 117
Asoka, 58
Assam, 26, 30, 39, 71, 77, 78, 92, 97, 139, 142, 198, 209
Atjeh, 236, 237
Attock, 137, 138

Aurangzeb, 61, 62
aus crops, 121, 141

BABUR, 61, 136
Balawan, 237, 239
Balikpapan, 239
Balkh, 58
Baltistan, 145, 146
Baluchistan, 44, 56, 128, 129, 130, 134, 135
Banares, 104, 114
Bandjermasim, 239
Bandoeng, 34
Bangalore, 119
Bangkok, 39–44, 218, 222, 231, 233
Banihal Pass, 146
Banka, 32, 237, 239
Bannu, 137, 138, 139
Barakar river, 95
Barind, 141
Barisal, 38
Batalagodawewa tank, 150, 152, 153
Batticaloa, 150, 152, 153
bauxite, 118
Bawdwin mines, 223
Beas river, 57, 105, 129, 133, 134
Bengal, 15, 78, 80, 86, 87, 98, 100, 209, 219, 223, 243
Bengal, Bay of, 33, 35, 38, 39
Bengal plain, 28
Bengali, 16
beri beri, 70
Betwa river, 104
Beveridge, Lord, 89
bhadai harvest, 112
Bhadravati, 95, 96, 102
Bhakra-Nangal scheme, 83, 101, 103, 110, 111
bhangar, 107, 108, 110, 112, 129, 131, 141
Bhilai, 96, 102
Bhutan, 126, 127
Bihar, 76, 87
Bikaner, 116
Billiton, 32, 239
Bindusara, 58
Bokaro coalfield, 96

Bolan Pass, 130, 135
Bombay, 35, 39–44, 45, 71, 80, 86, 87, 90, 92, 94, 100, 120
Borneo, 18, 22, 30, 47, 52, 68, 74, 80, 217, 218, 234, 239, 240
boro crops, 141
Brahmaputra river, 24, 38, 90, 92, 121, 123, 139, 144
Broach, 85, 119
Brunei, 217, 239, 240
Buck, Professor J. L., 174, 175
Buddhism, 16, 58, 67
Buddhists, 16, 60, 100, 126, 146, 154, 219
Bukit Asem, 236, 237
burghers, 154
Burma, 16, 18, 19, 26, 28, 30, 32, 35, 39, 45, 47, 48, 49, 66, 71, 72, 73, 75, 77, 80, 157, 189, 209, 217, 219–25
Burma Road, 188
Burma-Siam railway, 231
Burmah Oi! Company, 220
Butterworth, 226

CAGAYAN valley, 241
Calcutta, 18, 35, 39–44, 45, 71, 81, 87, 90, 92, 94, 95, 105, 122, 198, 222
Calicut, 61, 68, 92, 120
Cambay, Gulf of, 56
Cambodia, 67, 75, 82, 217, 230–3
Cameron Highlands, 45
Canton, 166, 167, 171, 187, 188
Cantonese, 16
caravels, 68
Cardamom Hills (India), 114, 118, 119
Cardamom Hills (Indo-China), 232, 233
cassava, 77
casuarina trees, 49
Cauvery river, 97, 116, 119, 120
Cawnpore, 94, 112
Celebes, 27
Ceylon, 18, 19, 28, 30, 32, 35, 38, 45, 50, 68, 71, 73–5, 78, 80, 82, 148–63, 210, 211
Chagai Hills, 135
Chainat barrage, 229, 230
Chaine Annamitique, 232, 233
Chakraborty, Dr., 121
Chalna, 140, 144
Chaman, 130, 135
Chambal river, 104–8, 116
Champa rice, 65, 66
Chandernagore, 18, 89
Chandigarh, 110, 111, 133
Chandragupta I, 58
Chandragupta Maurya, 58
Chang Tang, 197, 198
Changchun, 195

Chao Praya—see Menam Chao Praya
Charnock, Job, 122
chars, 139
Chatra gorge, 127
Chauk, 220, 221
chena, 159, 160
Chenab river, 129, 131–3
Chengchow, 182
Chengtu, 185, 186
Cheribon, 235, 236
Cherrapunji, 39–44
Chhattisgarh, 115, 118
Chiang Kai-shek, General, 164
Chiang Mai, 39–44, 230, 231
Chihli, Gulf of, 28, 63
Chin Hills, 222
China, 19, 20, 30, 33, 35, 38, 39, 45, 47, 50, 53–5, 63–7, 68, 75, 76, 80–3, 164–202, 204, 217–19
China Proper, 164, 165, 167, 171, 174, 188, 192, 193
Chindwin river, 217, 220, 221, 223, 231
Chinese in South-east Asia, 217–19
Chinese People's Republic, 125
Chinghai, 191, 197, 198
Chinwangtao, 182
Chittagong, 38, 90, 122, 123, 128, 139, 140, 142, 143
cholera, 18, 87
Cholon, 233
chos, 108, 110
Chota Nagpur, 115, 117
Chungking, 166, 171, 185, 188
cinnamon, 157
citronella, 157
Clive, Robert, 68, 86, 87
Cochin, 120
cocoa, 157
coconut, 71, 78, 79, 155, 157, 225
coffee, 78, 97, 156
Coimbatoire, 119
coir, 156
Colombo, 34–46, 81, 150–5, 158, 162
Colombo Plan, 17
Confucianism, 16
copper, 125, 170, 189, 208, 242
copra, 156
cotton, 71, 90
Cressey, Professor G. B., 185
'culture and consignment' system, 234
Cuttack, 69

DACCA, 39–44, 140, 143, 144
Dairen, 167, 194, 195
Dalai Lama, 198, 199

Damão, 18, 19
Damodar river and hydro-electric scheme, 83, 95, 101, 103, 117, 122
Damodar Valley Corporation, 83, 117
Darjeeling, 39-44, 90, 122, 125
Deccan, 22. 51, 55, 58, 60-2, 70, 90, 92, 115
Deccan Traps (lavas), 22, 115, 116, 118, 119
Delhi, 35, 39, 45, 60-2, 71, 87, 94, 104, 105, 106, 110
Dhubri, 123
Digboi, 123
Dikoya, 158
disease, 18
Diu, 18, 19
doabs, 108, 131
Dobby, Professor E. H. G., 225, 241
Drug, 95, 96
Dry Zone of Burma, 217, 219-21, 223
Dry Zone of Ceylon, 148, 149, 151, 152, 158-62, 174
duns, 124
Durand, Sir Mortimer, 136
Durand Line, 136
Durgapur, 96, 102
Dutch, 18-20, 154, 155, 219, 233, 234, 241
Dyaks, 240
dysentery, 18, 87
Dzungarian Basin, 20, 190, 191, 199

EARTHQUAKES, 23, 27, 178
East India Company, 68, 88, 122
East Nara river, 129
East Pakistan, 16, 52, 123, 128, 139-44
Emmahaven, 236
estate agriculture, 78, 79, 97, 156, 157, 162
Everest, 25, 114

FAR EAST, 21
Farmer, Mr. B. H., 160
Fatehpur Sikri, 112
Federation of Malaya (see also Malaya), 218, 228, 240
Fen Ho, 178
Fengfeng, 180, 182
Ferghana, 61, 191
Fisher, Professor C. A., 212
Foochow, 39-45, 187
Formosa—see Taiwan
Fort Sandeman, 135
Fort William (Calcutta), 87
France, 18
French, 19, 217
Fukien, 16
Fukienese, 16
Fushun coalfield, 167, 168, 195

GAL Oya, 150, 161
Galle, 150, 158
Gandak river, 105
Gandhi, 98
Ganges delta, 39, 71, 121
Ganges plains, 35, 38, 39, 51, 61, 69, 70, 75, 86, 87, 90, 97, 104, 107, 109-13, 122, 135, 139, 147
Ganges river, 23, 29, 35, 58, 105, 144
Ganges/Kobadak irrigation scheme, 143
Gauhati, 123
Gaya, 114
Genghis Khan, 66
Georgetown, 226, 227
Ghaggar river (or Sarasvati), 56, 108, 110, 116, 129
Gilgit, 145, 146
Goa, 18, 19, 85, 120
Goalundo, 140, 144
Gobi Desert, 20, 35, 47, 165, 167, 190, 192, 196
Godavari river, 97, 116, 120
Godwin Austen mountain, 24
Gogra river, 105, 109, 110
Grand Canal, 64, 65
Great Khingan mountains, 28, 195, 196
Great Wall of China, 63, 65, 70, 164, 165, 176, 190, 192
Greater East Asia Co-prosperity Sphere, 209
Greater Malaysia, 240, 241, 243
Gujarat, 119
Gujranwala, 134
Gupta empire, 58-60
Gurkha Rifles, 127
Gwalior, 116

HAINAN, 170, 188
Halmahera, 27
Han Dynasty, 63, 64, 204, 213, 218
Hangchow, 66, 167
Hankow, 184
Hanoi, 188, 189, 233
Hanyang, 184
Harappa, 56
Harbin, 194, 195
Hardwar, 112, 113
Harnai Pass, 130, 135
Hatton, 158
Hawkins, William, 68, 85
Hazaribagh, 118
High Asia, 17, 20
Himalayas, 23, 26, 47, 48, 50, 104, 144-6
Hindu Kush, 23, 24, 27, 61, 135
Hinduism, 16, 58, 114
Hindus, 16. 60, 98, 99, 104. 128, 146, 154
Hirakud hydro-electric scheme, 103

Hokkaido, 70, 203, 204, 206, 207
Holland, 19
Hong Kong, 18, 45, 80, 82, 187, 188, 200–2, 209
Honshu, 203, 204, 206, 207
Hooghly river, 121, 139
Hooghlyside, 122
hot-weather depressions, 38, 39
Howrah, 122
Huhehot, 191, 196
Hukawng valley, 221, 223
Hwainan, 184
Hwang Ho, 51, 64, 164, 175, 176, 178, 179, 181, 183, 195, 196
Hwang Ho hydro-electric scheme, 83, 169, 176
Hyderabad (Deccan), 88, 118
Hyderabad (Sind), 129, 130
hydro-electric schemes, 83, 84, 102, 103, 119, 120, 127, 162, 169, 189, 195, 207, 208, 214

ICHANG, 39–45
indaing, 48
India, 18–20, 28, 30, 47, 54, 61, 67, 70, 72–5, 80–3, 95–127, 157, 243
Indian National Congress, 98
Indo-China, 18, 19, 22, 29, 38, 48, 217, 218, 231–3
Indo-Gangetic plains, 29, 146
Indonesia, 16, 18–20, 33, 38, 67–9, 75, 77, 210, 218, 219, 233–9
Indus civilization, 56, 57, 106, 174
Indus plain, 28, 47, 56, 61, 75, 90, 128, 129
Indus river, 29, 56, 57, 97, 105, 133, 134, 139, 147, 197
Inland Sea, 203, 207
Inner Mongolia, 192, 195, 196
Inner Tibet, 192, 197–9
Ipoh, 225, 227
Irian, 241
iron and steel, 80, 90, 95–7, 102
Irrawaddy river, 25, 28, 29, 53, 216, 217, 220–3, 231
irrigation, 72, 74, 75, 83, 87, 97, 101, 103, 106, 109, 130–4, 159, 220, 229, 243
Islam, 16, 17, 176
Islamabad, 138

JAFFNA, 149, 150, 152, 153, 155, 157, 158, 161, 162
Jainism, 58
Jaipur, 116
Jakarta, 34, 219, 235
Jammu, 145, 146

Jamshedpur, 95, 96, 102, 120
Jamuna river, 139, 140
Japan, 16, 19, 23, 27, 30, 33, 35, 38, 39, 50, 54, 68, 73–5, 80–3, 202–12
Java, 23, 27, 30, 35, 47, 53, 67, 72, 78, 80, 217, 219, 233–6
Jesselton, 240
Jharia coalfield, 95, 96
Jhelum, 134, 135
Jhelum river, 57, 129, 132–4, 145
jhum, 143
Jinnah, Muhammad Ali, 98
Jodhpur, 116
Johore, 226–8
Jullundur, 110
Jumna river, 105, 106, 109, 110, 112
jute, 71, 80, 90, 94, 95, 99, 100, 120, 142, 187, 222

KABUL, 61, 136–8
Kabul river, 136–8
Kai Tak airport, 201, 202
Kaifeng, 179, 182
Kailan Basin, 182
Kailas mountains, 24
Kaili, 167
Kalabagh, 134
Kalat, 130
Kalimantan, 239
Kalimpong, 122, 198, 199
Kanchenjunga, 24, 25, 127
Kandahar, 135, 136
Kandy, 150, 155, 158
Kansu, 168, 175, 176, 191
Kanto plain, 203
kaoling, 70, 176, 178, 179, 193
Karachi, 39–44, 80, 100, 128–31, 134, 135
Karakoram, 23, 24, 144–6
Karens, 217
karez, 72, 134, 136
Karikal, 89
Karnafuli river, 140, 143
Kashgar, 46, 146, 199
Kashmir, 100, 138, 144–7
Kathiawar, 59, 90, 105, 119
Katmandu, 126
Kelantan delta, 225–7
Ken river, 104
Kerala, 120
Kerintje mountain, 236, 237
khadar, 107, 108, 110, 112, 129, 131
kharif crops, 110, 112, 131
Khmers, 232
Khotan, 145, 146, 199
Khulna, 140, 143
Khyber Pass, 135–7

Kialing river, 184, 186
Kiaolai corridor, 181, 182
Kirana Hills, 131
Kirin, 195
Kirthar mountains, 23, 129
Kistna river, 70, 97, 103, 116, 118, 120
Kobe, 205, 207
Kohat, 137–9
Kolar goldfields, 119
Korat, 28, 229, 231
Korea, 28, 209–15
Kosi river, 105, 114, 124, 127
Kota Bharu, 39–44, 227
Kotri Barrage, 130, 131
Kowloon, 200–2
Kra Isthmus, 223
Krakatau, 27
Kuala Lumpur, 225, 227
Kuantan, 226, 227, 229
Kublai Khan, 66
Kuching, 240
Kun Lun mountains, 24, 145
Kunar river, 136, 137
Kunming, 186, 188, 223
Kwangsi, 170
Kwanting reservoir, 181
Kwantung peninsula, 194
Kweichow, 167, 187–89
Kyushu, 203, 204, 206

LADAKH, 145, 147
Lahore, 39–45, 131, 134
Lamaist Buddhism, 165, 198
Lanchow, 20, 165, 176, 182, 197
lantana scrub, 49, 50, 53
Laos, 217, 231–3
Lashio, 188, 189, 223
laterite, 52
laterization, 52, 53
Lhasa, 35, 67, 122, 125, 127, 165, 167, 171, 192, 198, 199
Lhodran, 134
Liaoning plains, 193
Liaotung peninsula, 28
Linfen, 167, 178, 182
loess, 53, 174–9
Lothal, 56
Loyang, 182
Luang Prabang, 230
Lucknow, 104, 112
Ludhiana, 110
Lunghai railway, 182
Luni river, 116
Luzon, 241, 242
Lyallpur, 132, 134

MAANSHAN, 167
Macao, 18, 68, 202
Madhapur Tract, 141
Madoera, 233–5
Madras, 35, 39–44, 49, 60, 61, 86, 92, 94, 120
Madura, 119, 120
Magadha, 58
Mahanadi river, 69, 103, 116, 118, 120
Maharashtra, 115, 118
Mahaweli river, 150
Mahé, 89
Majapahit, 67
Makran coast, 134, 135
Malabar coast, 34, 68
Malacca, 227, 228
Malacca Strait, 67
Malakand hydro-electric scheme, 138
Malakand Pass, 137, 138
malaria, 18, 87, 148, 149, 161
Malaya, 16, 18, 19, 22, 28, 32, 35, 38, 46–9, 52, 54, 68–71, 73, 75, 78, 82, 83, 169, 209, 210, 217, 218, 225–9, 242
Malwa, 115, 116
Manchu Dynasty, 66, 198
Manchukuo, 195
Manchuria, 28, 30, 50, 66, 70, 165, 167–71, 190, 193–5, 209, 210, 213, 214
Manchus, 66, 67, 182, 189, 192, 196, 198, 213
Mandalay, 39–45, 53, 220, 221, 225
Mandarin, 16
Mandi, 134
mangrove, 31, 49, 50, 224, 226, 236, 238
Manila, 80, 241
Manipur, 123
Maratha states, 87
Marathas, 86, 100, 118
Matsu, 188
Mauryan empire, 57, 58
Mayurbhanj, 95, 96
Medan, 236, 237, 239
Meerut, 104
Meghna river, 139, 140
Mekong development scheme, 233, 243
Mekong river, 25, 197, 216, 223, 232, 233
Menam Chao Praya, 29, 216, 229–32
Merapi mountain, 27
Merauke, 241
mercury, 189
Mergui, 39–44, 221, 223
Mettur dam, 119
Middle East, 17
Min river, 184–6
Mindanao, 241, 242
Mindanao trough, 27
Mindoro, 241
mineral resources, 29–32
Ming Dynasty, 66, 193

Minneriya Development Company, 160
Minneriya tank, 159
Miri, 240
Moesi river, 236
Moghul empire, 61, 62
Mohenjo-daro, 56
Mongol Dynasty, 66
Mongolia, 20, 27, 66, 192
Mongolian Autonomous Ch'u, 195
Mongolian People's Republic, 20, 191, 192, 196, 197
monsoon, 20, 33–44 (see also winter monsoon and summer monsoon)
Monsoon Asia; boundaries, 19, 20; suitability of name, 20, 21
Montgomery, 132, 134
Monywa, 220
Moors, 154
Moulmein, 49, 216, 221, 223
mud volcanoes, 26
Mukden, 167, 171, 195, 214
Multan, 134, 135
multi-purpose hydro-electric schemes, 83, 84, 102, 103
Muri, 118
muslims, 16, 55, 60, 61, 64, 71, 98, 99, 128, 135, 136, 142, 146, 157, 165, 176, 199, 239
Musoorie, 125
Mysore, 58, 60, 88, 97, 115, 118, 119

NABOBS, 87
Naf river, 224
Nagasaki, 205, 207
Nagoya, 205, 207
Nagpur, 94, 118
Nakhaun Sawan, 229, 230
Namcha Barwa mountain, 24
Nan Shan (Central Asia), 24, 30, 168, 190, 191
Nan Shan (South China), 28
Nancheng, 167
Nanga Parbat mountain, 24, 144
Nani Tal, 125
Nanking, 167, 170, 184
Narbada river, 59, 61, 115, 119
nationalism, 18, 19
Naula Tank, 150, 152, 153
Negrais, Cape, 222
Negros, 241
Nepal, 114, 126
Netherlands East Indies, 18, 156, 217, 234
New Guinea, 19, 20, 27, 29, 74, 217, 241
Nias, 27
Nicobar Islands, 26
Nilgiri Hills, 97, 118, 119
Noakhali, 38

North Borneo, 217, 239
North China plain, 29, 174, 179–83, 193, 206
North Karanpura coalfield, 96
North Vietnam, 30, 232, 233
nullahs, 116
Nuwara Eliya, 150, 152, 153

OIL palm, 78, 79, 81, 225, 239
oil refining, 80, 222
oilfields, 29–31, 83, 168, 176, 207, 220, 221, 234, 236–41
Ootacamund, 119
Ordos plateau, 195
Orient, 21
Osaka, 205, 207
Outer China, 164, 165, 167, 190–200
Outer Mongolia, 192
Outer Tibet, 192

PADANG, 236
padauk, 48
Padma river, 139, 140, 143
pagodas, 219
Pahang-Rompin valleys, 226, 227
Pakhtunistan, 138, 139
Pakistan, 18, 19, 47, 70, 71, 73, 75, 81, 82, 98–100, 123, 128–47, 243
Palawan, 30, 241
Palembang, 236
Palk Strait, 157, 163
Pamirs, 20, 24, 27, 59, 66, 165, 190
Panay, 241
Pangkalanbrandan, 238
Pangkalansoesoe, 237, 238
Panipat, 61
Paoki, 167, 178, 182, 186
Paotow, 166, 167, 169, 170, 196
Parsees, 100, 128
Pataliputra, 58, 114
Pathans, 136, 138, 139, 146
Patna, 58, 104, 114
Pearl Harbor, 209
Pedurutalagala mountain, 149
Pegu Yoma, 220
Peking, 20, 29, 35, 39–44, 66, 164, 167, 169, 171, 181, 182
Penang, 39–44, 225–8, 239
Peradeniya, 158
Periyar river, 119
Peshawar, 136–9
Philippines, 19, 23, 27, 30, 33, 35, 38, 48, 68, 75, 77, 169, 217–19, 241, 242
Phnom Penh, 230, 231, 233
Pingsiang, 167, 184
Pir Panjal mountains, 144, 145

plague, 18
plantation agriculture, 78, 79, 97, 156, 157, 162
plumbago, 151
Polonnaruwa, 150, 159, 160
Pondicherry, 18, 86, 89, 120
Pontianak, 45, 239
Poona, 118
Popa mountain, 222
Port Arthur, 194, 195
Port Dickson, 227
Port Okha, 119
Port Swettenham, 225–8
Port Weld, 225–7
Portuguese, 18, 68, 85, 86, 157, 217
Potwar plateau, 129, 137, 138
Poyang, 183, 186
Prai, 226, 227
Punjab, 38, 56, 58, 61, 70, 71, 87, 92; partition of, 97–100, 131, 132, 134, 144, 147
Pushtu, 136
Puttalam, 150–3, 158
pynkado, 48
Pyongyang, 213

QUANG Tri, 39–44
Quang Yen coalfield, 230, 233
Quemoy, 188
Quetta, 23, 130, 135
quinine, 156

rabi crops, 110, 112, 131, 143
rabi harvest, 114
Raffles, Sir Stamford, 228
Rai Pithora, 105, 106
Rajasthan, 116
Ranchi, 118
Rangoon, 26, 30, 39–45, 49, 219, 220, 222, 225, 266
Raniganj coalfield, 95, 96
Rann of Cutch, 116, 119, 129
Ravi river, 129, 131, 132
Rawalpindi, 57, 138, 146
Red river, 29, 188, 216, 219, 232
regur, 116, 118
Roe, Sir Thomas, 85
Rourkela, 96, 102
rubber, 71, 78, 79, 81, 97, 155, 157, 158, 162, 218, 223, 225, 226, 228, 231, 232, 234, 239, 240, 242

SABARMATI river, 119
Sahul Shelf, 28
Saigon, 233

Sailendra kings, 67
sal, 48
Salem, 119
Salt Range, 116, 128, 129, 134, 137, 138
Salween river, 25, 188, 197, 216, 223
Sambhar lake, 116
Sandakan, 240
Sanmen Gorge, 169, 181
Sarasvati (or Ghaggar river), 56, 108, 110, 116, 129
Sarawak, 217, 239, 240
Satpura Hills, 115, 118
Sawahloento, 236, 237
Sayan mountains, 28
Seleukos Nikator, 57
Seoul, 213, 214
Shabad, Theodore, 187
Shahjahan, 61, 105, 112
Shahjahanabad, 105
Shan plateau, 28, 45, 223
Shan States, 30, 49
Shanghai, 166, 167, 170, 178, 183, 188
Shans, 217
Shansi, 167, 168, 178
Shantung peninsula, 28, 51, 167, 174, 179, 181, 182
Shensi, 167, 168, 176, 178
Shi Huang Ti, 63
shifting cultivation, 77
Shihkingshan, 167, 182
Shihtsuishan, 167
Shikoku, 203
Shillong plateau, 28, 123
Sholapur, 94, 118
Siam—see Thailand
Sian, 64, 167, 178
Siangtan, 167
Siberut, 27
Sibolga, 236, 237
Sihanoukville, 230, 233
Sikang, 186, 191, 197, 198
Sikhs, 98, 99
Sikiang river, 63, 165, 187, 202
Sikkim, 24, 125, 127
Simeuloee, 27
Simla, 39–44, 125
Simla Conference, 199
Sind, 35, 44, 87, 97
Sindri, 117
Singapore, 19, 39–46, 69, 81, 82, 209, 218, 219, 225–9
Singhbhum, 96
Singkawan, 239
Sinhalese, 154, 155, 157, 160, 163
Sinkiang, 30, 165, 167, 168, 176, 182, 191, 192, 199, 200
Siri, 105

Sittang river, 28, 29, 216, 220–2
Siwalik Hills, 104, 105, 109, 113, 124, 144, 145
Solo river, 235
Son river, 58, 104, 105, 114
Soochow, 167
South Vietnam, 232, 233
South-east Asia, 21, 66, 67, 69, 77, 83, 86, 157, 209–11, 216–42, 243
Spain, 19
Spate, Professor O. H. K., 104, 120, 136
Srinagar, 144–6
Srivijaya, 67
Stamp, Professor L. D., 203
steel, 80
Subaranekha river, 95
Sui, 128, 131, 134, 135
Sukkur Barrage, 97, 129, 130, 134
Sulaiman mountains, 23, 134
Sultanate of Delhi, 60
Sumatra, 23, 26, 27, 29–31, 47–9, 67, 74, 78, 80, 216, 217, 234, 236–9
summer monsoon, 33–5, 39–46, 189
Sun Yat Sen, 164
Sunda Platform, 28, 30, 32, 52, 223, 239
Sunda Strait, 27, 236
sundarban, 121, 139, 140
Sung Dynasty, 64–6
Sungari river, 169, 194, 195
Surabaya, 39–44, 235, 236
Surat, 85, 86, 105, 119
Surma river, 139, 140
Sutlej river, 83, 101, 103, 108, 109, 110, 129, 132
Swat, 136–8
Sylhet, 139–44
Syr Darya, 61
Szechwan, 30, 167, 168, 170, 174, 184–6, 197

Tachin river, 229, 230
Taihang mountains, 28, 29
Taipeh, 189
Taiping, 225, 227
Taiwan, 75, 82, 164, 168, 170, 188–90, 209–12
Taiyuan, 167, 178
Taj Mahal, 112
Takao, 189
Taklamakan Desert, 47, 165, 190
Tamils, 154, 157, 158, 163
Tang Dynasty, 64, 65, 206
Tangpu, 182
Tangshan, 182
Tanjore, 61
tanks, 72, 148, 159, 160, 162
tanr crops, 117

Taoism, 16
Tapti river, 115, 119
Tarakan, 239
Tarim Basin, 20, 23, 24, 27, 47, 59, 63, 64, 70, 72, 190, 191, 199
Tata, J. N., 95
Tatung, 176
Taunsa Barrage, 134
Tavoy, 221, 223
Taxila, 57, 58
tea, 78, 79, 97, 142, 156, 157, 162, 187, 232, 239
teak, 48, 49, 222, 223, 225, 229–31
Telok Anson, 226, 227
Tennasserim, 221, 223, 225
terai, 124, 127
Tezpur, 123
Thailand, 16, 19, 22, 29, 32, 48, 68, 70, 71, 75, 217, 218, 225, 229–31
Thais, 217
Thal Project, 134
Thar Desert, 104, 114, 116, 129
Theyetmyo, 220, 221
Tibet, 20, 24–6, 35, 38, 46, 47, 125, 127, 164, 165, 188, 190–2, 197–9
Tibet Autonomous Ch'u, 199
Tien Shan, 23, 24, 30, 168, 199
Tientsin, 167, 171, 179, 181, 182
Tihwa, 168, 171, 199, 200
Timor, 18, 27, 48, 217
Tin, 30, 32, 189, 218, 223, 225, 226, 228, 231, 234, 239
Tista river, 24, 127, 140
Tjilatjap, 235
Tokyo, 23, 39–44, 203, 206, 207
Tonle Sap, 230, 232, 233
Trans-Mongolian railway, 196
Travancore, 97
Trengganu highland, 226, 227
Trichinopoly, 61, 119
Trincomalee, 150, 151, 157, 162
Triple Canal Project, 131–3
Tripura, 123, 140
Truman, President, 214
Tsaidam depression, 191, 197
Tsangpo river, 197
Tsinan, 167, 181, 183
Tsingtao, 166, 167, 171, 181, 182
Tsining, 196
Tsin-ling Shan, 25, 28, 35, 45, 171, 174, 176
Tsunyi, 167
tube wells, 72, 97, 109
Tughlaqabad, 105, 106
Tumen river, 213
tung oil, 78, 184, 187, 223
Tungabhadra hydro-electric scheme, 103, 118

Tungans, 199
tungsten, 170, 213
Tungting, 183, 186
Turfan, 28, 199
typhoons, 34, 38, 39, 44, 46

UIGHERS, 165, 199
Ulan Bator, 46, 196
United Nations, 17, 73, 74
uranium, 84
Urdu, 16
Urumchi—see Tihwa
U.S.A., 19, 102, 242
U.S.S.R., 17, 20, 24, 102, 165, 170

VASCO da Gama, 68
Vientiane, 230
Vietnam, 83, 217, 219, 231, 232
Vijayanagar, 60–2
Visayan Islands, 241
Vizagapatam, 120
volcanoes, 27

WARSAK, 138
Watawala, 39–44
Wei Hai, 180, 182
Weihaiwei, 182
Weihsin, 181
West Irian, 241
West Nara, 129

West Pakistan, 16, 30, 72, 80, 128–39
Western Ghats, 28, 48, 86, 118
Wet Zone of Ceylon, 148, 149, 151, 152,
    155–8, 160, 161
Whampoa, 187
wheat, 70
Wingate, General Orde, 26
winter depressions, 34, 38–44, 46
winter monsoon, 33, 35, 37–44, 46
World Bank, 17
Wuchang, 184
Wuchow, 187
Wuhan, 166, 170, 184, 188
Wuhu, 184

YALU river, 169, 213, 214
Yanaon, 89
Yangtze river, 25, 44, 64, 165, 167, 168, 170
    183–5, 197
Yangtze valley, 38, 44, 45, 51, 69
Yarkand, 199
Yenangyat, 220, 221
Yenangyaung, 220, 221
Yokohama, 207
Younghusband, Sir Francis, 127, 198
Yukikow, 184
Yummen, 30, 168, 172, 176, 184
Yungting, 181, 182
Yunnan, 25, 28, 30, 50, 167, 170, 188, 223

ZOB valley, 135

9 780202 309422